The future of DNA

The future of DNA

Proceedings of an international *If*gene conference
on presuppositions in science and expectations in society
held at the Goetheanum, Dornach, Switzerland,
2nd – 5th October 1996

Edited by

J. WIRZ
Forschungsinstitut am Goetheanum, Naturwissenschaftliche Sektion,
Dornach, Switzerland

and

E.T. LAMMERTS VAN BUEREN
Louis Bolk Institute, Driebergen, The Netherlands

SPRINGER SCIENCE+BUSINESS MEDIA, LLC

A C.I.P. Catalogue record for this book is available from the Library of Congress

ISBN 978-94-010-6312-8 ISBN 978-94-011-5494-9 (eBook)
DOI 10.1007/978-94-011-5494-9

Cover photograph: © Professor Peter Baude, Embryo and Gamete Research Group

Table of contents

The Goetheanum

Preface

This book summarizes the efforts and results of the first international *If*gene conference on presuppositions in science and expectations in society with respect to genetic engineering which was held at the Goetheanum, Dornach, Switzerland, October 2-5, 1996.
The Goetheanum provided a unique opportunity to gather people from diverse disciplines who have opposing attitudes on modern science and technology. It is due to this venue, among other things, that the participants were able to develop an open, power-free dialogue and could focus more on judgement-forming than a polarizing debate.

This *Ifgene* conference could not have happened without the financial support from many private individuals and the following organisations listed in no particular order: Fetzer Foundation; Stichting Triodos; Evidenzgesellschaft; Mahle Stiftung GmbH; Gemeinnützige Treuhandstelle e.V. Bochum; Initiative gegen 'Bioethik'; Verband für anthroposophische Heilpädagogik – CH; Verband für anthroposophische Heilpädagogik, Sozialtherapie und Sozialarbeit e.V. – D; Stichting ter bevordering van de Heilpaedagogie; Iona Stichting; Antroposofische Vereniging in Nederland; Stichting Klaverblad; Swissair & Crossair; The Rudolf Steiner Association; The Welcome Association; Anthroposophische Gesellschaft in Deutschland; Helixor Heilmittel GmbH & Co; Goetheanum Dornach; Verein für anthroposophisches Heilwesen e.V. – D; The Oakdale Trust; Unilever Nederland BV; NV Verenigde Bedrijven Nutricia; Migros-Genossenschafts-Bund; Ministerie van Landbouw, Natuur en Visserij; CIBA; Stichting Elise Mathilde Fonds; Anthroposophic Society Australia; Therexsys; COOP Schweiz and selected speakers (G. Stotzky – USA, H.A. de Boer – NL, J. Girard-Bascou – F, P.G. Middleton – AUS, N.G.J. Jaspers – NL) were supported by the Swiss Academy of Sciences (SANW).

This publication has benefitted from the concerted action of all contributors, the critical and patient editing by David Heaf, Pat Cheney and John Armstrong, the editorial help of Saskia Kneulman, Henk Pel, and the layout by Harm van der Meulen.

We hope that this book provides a contribution to and enlargement of the ongoing scientific, social and philosophical discussions on biology and genetic engineering.

Johannes Wirz and Edith Lammerts van Bueren

Introduction

John Armstrong, Christine Ballivet, David Heaf, Manfred Schleyer, Meinhard Simon, Edith Lammerts van Bueren, Johannes Wirz

Ifgene coordinators

Genetic engineering is increasingly becoming part of our daily lives. For instance, the food processing industry depends on it to a large extent and many modern diagnostic test in medicine are based on methods derived from DNA technology. Along with these advances, the public is becoming more aware of the enormous potential of the technology, as well as the ethical and social issues related to it. Since the conference has taken place, two events have continued to sharpen public awareness and to deepen the controversy: the arrival of transgenic soya beans in Europe, as well as the first, successful cloning of a mammal from tissues of an adult animal, the lamb Dolly.

Scientific views about DNA and genes challenge our fundamental concepts about life, nature, society and humanity. The public debate about genetic engineering is based on a paradigm that seems to be widely accepted by scientists, as well as by laymen. It is the paradigm of reductionist biology, which postulates that all attributes and characters of life in its substance and form are ultimately determined by genes. Other factors like the natural and the social environment are recognised as being only of secondary importance.

There are however other possible approaches to an understanding of life. Some of them stress the contextual and relational qualities of organisms and consider them to be the basic cause rather than the consequence of molecular interactions at the genetic, i.e. the DNA, level. They acknowledge that every living being is endowed with its own dynamics, sustained by the interaction of both environment and genes. But approaches to understanding life that encompass genetic determinism are also conceivable. Indeed, molecular biological discoveries themselves prompt us to search for such approaches.

Such a search would be of value not only to philosophers of science or epistemologists, but also to all those concerned with biological science and its application. From the outset, our concepts and ideas shape our perceptions of the world and determine our actions. Thus, ethical or moral values necessarily reflect our scientific outlook on the world.

Some initial questions related to the scientific and social aspects of genetic engineering can be identified: Where does the power of this technology originate from? What characters and properties of living beings does it unravel? Where and how does it come up against limitations?

A second group of issues relates to the presuppositions of DNA thinking. The success of molecular biology often hides the fact that its scientific and philosophical foundation is open to being questioned and reflected upon like any approach to understanding life. Obviously, such reflections are more fundamental than socio-economic interests and concerns, which are anyway to do with applications of the technology. Indeed they transcend an ethical debate which is restricted to risk-benefit assessment, be it in ecology, public health or social rights.

At the *If*gene conference the fundamental issues were tackled in several different ways. On the first day, the discussion focused on scientific and social aspects. The introductory lectures shed light on the benefits, challenges and dangers of DNA-thinking and faced the question: What will our world and society look like if they are shaped by the concepts of molecular genetics? What qualities of science and society will be deepened and enlarged by gene thinking? Which qualities would be lost and how can they ultimately be salvaged, reintroduced or formed anew?

The second day covered molecular genetics in biology. The rate of discovery of new genes and their functional properties and interactions is breathtaking. Our insight into molecular function is highly advanced and will develop in still greater depth. However, when molecular biology moves from a descriptive to an explanatory science, obstacles are encountered. Molecular function does not readily explain pattern formation during development or processes of consciousness etc. The fate of transgenic organism in the environment cannot be deduced from the results of DNA manipulation or calculated in advance. Thus, the theories based on the molecular approach fail to explain life-processes. Are there essential aspects missing?

The third day was dedicated to DNA and the human being. Faced with the serious issues about the social impacts of the new technology, public, scientific and medical awareness is severely challenged. Diagnosis and therapy open a whole field of new questions which require us to rethink and reformulate concepts such as human individuality, health and disease.

Participants in the evening round-table discussions shared their attitudes towards genetic engineering and the aspects of their personal biographies that led them to take their particular positions. These discussions showed that besides the ability to grasp certain 'objective' facts about this technology, the contextual environment, i.e. the 'personal subjective approach' is of equal importance for judgement formation.

The *If* gene conference brought together people – both scientists and non-scientists – who wanted to raise the dialogue to create a pluralistic exchange of concepts, hypotheses and pictures of what it is to be human and the nature of the world. The discussion focused on the presuppositions, as well as the consequences and perspectives of knowledge, in order to raise consciousness and provide a broad foundation for individual ethical judgement forming.

2 The limits of culture in biotechnology

Klaus Michael Meyer-Abich
Practical philosopher of nature
Kulturwissenschaftliches Institut, Wissenschafiszentrum Nordrhein-Westfalen
Postfach 320240, D-45246 Essen

Abstract
We are supposed to leave the world as if we had not been there, but how far may we go in changing this planet? Culture is basically the most specific human contribution to the history of Nature. One element of culture is to respect dignity. Can limits of culture be identified in the course of biotechnology?

Introduction
The European journey of exploration into a future which was unknown and which characterises the second millennium after Christ, now involves the whole of mankind. What began with the crusades, under the cloak of belief, continued with the exploration of the earth, which led to colonization and the present global economic order. The earth proved to be round, and scientific discoveries and technological progress continued. Scientific progress so far has led to an end in the limits to growth of the industrialized economy but not in science and technology itself. It may well be however, that as far as biotechnology is concerned the earth will prove to be round once more, in the sense that we finally get back to our starting point, namely to ourselves as we set out to explore the world. Will scientific progress stop here or will Man himself be drawn into it?

Mankind's European initiative has certainly not been to the benefit of everyone. Yet I do not think it was basically mistaken. In the evolution of mankind there may be transitions which correspond to an individual's growing up. Thus for a society as a whole, the time may come to step out from the protection of the parent and find one's own feet. I believe that the biblical myth of the Fall refers to such a moment; that mankind has not been expelled from paradise but has set out from there to be free to find its own place in the world. Historically, the Neolithic revolution may have been such a transition. It seems to me that in our millennium and particularly at the time of the European Renaissance, being at home in a closed world again was no longer enough and the challenge was to find one's own home in an open world. This was the copernican revolution; relinquishing geocentrism and trying to discover how we fit into the world if our place is not at the centre.

As far as I can see there is nothing to be said against setting out for freedom into an open world. Growing up, however, involves concern for people and is no longer a simple matter of being cared for by others. In this sense the biblical 'Fall', though not a fault in itself, implied the possibility of fault, as men were in future to be held responsible for their own actions. The implications of this growing up seem to have been neglected in modern times, when geocentrism has finally been replaced by anthropocentrism, i.e. one wrong answer by another wrong answer.

The original Neolithic idea of human settlement in Nature was agriculture, which means to cultivate or civilise a place. This includes clearance – cutting trees, for instance – to make space for human beings. It makes a difference, however, whether the intention behind settlement is to live *with* others, plants and animals, in the community of Nature of which we ourselves are part, or to live *on* others, considering the rest of the world as a bag of resources to fulfill human needs or aspirations. Anthropocentrism is the idea that *being* human means *having* the world as a resource. Molecular biology and biotechnology so far have been developed along these lines. There is, however, a basic inconsistency in anthropocentrism because as soon as we begin looking at human beings, we cannot be ourselves and have ourselves at the same time. In medicine it has been felt for a long time that a human being cannot be adequately treated as a biochemical system. Since in molecular biology Man himself for the first time becomes an object of science in the full sense, this involves a challenge to overcome anthropocentrism. The directing principle for an alternative development is *culture* in the sense of integrity in living with others.

In the history of mankind culture has often enough been based on repression. Even in Athens there were slaves. At our level of political culture, however, integrity can no longer condone repression. In my understanding, the idea of culture is to be human in living *with* others – things and beings – instead of having them as resources. To be with others does not hurt anyone's dignity, and this is what human beings demand for themselves. To observe another's dignity does not mean to stand back from any interference. The artist, for instance, who carves a sculpture out of a stone hurts it but must not hurt its dignity. He may even reinforce or advance its inherent value with an enhancement which otherwise perhaps would not have come about. Goethe might have considered this advancement as a 'Steigerung' (intensification). In my understanding, things and beings do not have individual value or dignity just as they happen to be for themselves but their individual nature is what they tend to be in Nature as a whole.

Cultural activities are defined as those by which the world as a whole is better off than if such activities had not been engaged in. Art and cultivated landscapes – including cities – are the most obvious examples. It seems that culture is the most specific human contribution to the history of Nature and that this is the way in which Mankind enhances the world, which is thus better for it.

Of course, we can almost never be certain whether or not any one of our human activities really can be considered culture. This uncertainty goes with practically every important decision in human life. Yet the criterion of culture is open to reason. My objective in this paper is to give a preliminary account of what the idea of culture would mean for the development of biotechnology. Since artists are generally better at determining culture as opposed to non-culture, I shall refer to an artist in my first section. There is one who was also a scientist and who still to the present day remains the greatest challenge to the anthropocentric mainstream in science, namely Goethe.

Culture in science

Goethe is to be credited with a number of scientific discoveries, particularly that of the os intermaxillare (intermaxillary bone) which proved the kinship between man and ape. More than that he suggested a paradigm in botany which sounds as if he would have been quite happy with the design of transgenic plants. As he recorded in Naples on May 17th, 1787; the primordial plant (Urpflanze) – which he was then exploring – "will be the oddest creature in the world, which Nature herself ought to envy me (das wunderlichste Geschöpf von der Welt, um welches mich die Natur selbst beneiden soll). With this model and the key to it one can invent plants endlessly which must be consistent, that is if they did not exist, yet they *could* exist, ... having an inherent truth and necessity (Mit diesem Modell und dem Schlüssel dazu kann man alsdann noch Pflanzen ins Unendliche erfinden, die konsequent sein müssen, das heißt, die, wenn sie auch nicht existieren, doch existieren könnten und ... eine innerliche Wahrheit und Notwendigkeit haben). The same law will be applicable to everything living (Dasselbe Gesetz wird sich auf alles übrige Lebendige anwenden lassen)" (HA XI 375). Wouldn't it be justified from this to honour Goethe as a father of biotechnology?

Goethe's presentient sympathy for the design of artificial organic beings seems to be confirmed by his written correspondence with the pharmaceutical chemist, H.W.F. Wackenroder. In a letter to him he declared his interest in "how one could discover the organic chemical operation of life which achieves the metamorphosis of plants (Es interessiert mich hochlich, inwiefern es möglich sei, der organisch-chemischen Operation des Lebens beizukommen, durch welche die Metamorphose der Pflanzen ... bewirkt wird)." This is a firm endorsement for research into the chemistry of life. In the same letter he advocated "so to speak to corner the incomprehensible (das Unbegreifliche gewissermaßen in die Enge zu bringen)." Man could in fact "not desist from the attempt to corner the incomprehensible until he contents himself in doing so and *willingly admits to be overcome* (so kann der Mensch ... doch nicht von dem Versuche abstehen das Unerforschliche so in die Enge zu treiben, bis er sich dabei begnügen und sich willig überwunden geben mag)" (January 21, 1832 = HAB IV 467f.; my italics).

Apart from the last few words the argument sounds familiar from the current understanding in science, namely: biochemistry is the right way to understand life and we must embark upon what is not yet understood. But then the declaration would continue: ... upon what is not yet understood until *it* is overcome, while Goethe considered the opposite to be the case, that in the end *we* were to be overcome. Obviously, it makes quite a difference whether in tackling Nature scientifically ultimately she is overcome or we are overcome. The ruling paradigm in modern science definitely is that we are to overcome Nature, not the other way around.

What was Goethe's idea in letting ourselves be defeated by the object of knowledge? He surely didn't mean that it would be better if the wolf killed me rather than that I should kill the wolf, assuming for the sake of simplicity that there is no third case. An answer is found in Goethe's famous conversation with Schiller in July 1794. As Goethe reported, he vividly presented to Schiller how he experienced the metamorphosis of plants "letting a symbolic plant develop itself before my eyes (und ließ ... eine symbolische Pflanze vor seinen Augen entstehen)". Schiller listened kindly but finally shook his head and said: "That is no experience, that is an idea (Das ist keine Erfahrung, das ist eine Idee)." Goethe was annoyed but finally replied: "So I am lucky to have ideas without knowing it and even to perceive them with my eyes (Das kann mir sehr lieb sein, daß ich Ideen habe ohne es zu wissen, und sie sogar mit Augen sehe)" (HA X 540f.).

What Schiller as a Kantian considered to be an idea, was a human design to regulate experience according to the regime of our consciousness. In Kant's critical philosophy the laws of Nature were supposed to be prescriptions by the human mind and not the order of Nature itself, i.e. the world is as it is because we are as we are. Knowledge of Nature then becomes a matter of knowing oneself. This is the common denominator of German idealism in philosophy as well as of Romanticism in poetry. And there was nothing in the world Goethe felt to be more disgusting than this subjectivism of modern times. He certainly appreciated becoming aware of himself through critical and idealistic philosophy but his objection was that this approach "never reaches the object (sie [die kritische und idealistische Philosophie] kommt aber nie zum Objekt)." The object's existence, however, "we, like ordinary common sense, have to admit and to hold faithfully to it, if we are to enjoy life (dieses müssen wir so gut wie der gemeine Menschenverstand zugeben, um am unwandelbaren Verhältnis zu ihm die Freude des

Lebens zu genießen)" (September 18, 1831, to Schultz: HAB IV 450). Before this the old man – fifteen years after Christiane's death – had described how the ordinary people enjoyed life in recognizing 'the object' when they fetched water from the water pipe in front of his house. Compared to that, idealism as well as romantic poetry had only intellectual delusions to offer.

In acknowledging the object in human knowledge, Goethe rejected the alleged modesty by which Kant had raised the human subject higher than it had been conceived ever before (cf. HA X 539). Kant taught one to clip one's wings and reconcile oneself to conscious subjectivity because human reason comprehended nothing but its own products. Though Kant himself gave up this idea in his postcritical philosophy it has become normative in industrial society. In fact we have come to recognize only what we have produced or could have produced ourselves. This outlook shows what is at stake in the controversy between Goethe and Kant when the implications are considered with respect to the organisation of living beings. Goethe considered, as opposed to Kant, that rather than restricting human reason to its own design and construction, it is appropriate that one should release oneself from the narrowness of subjectivity.

While Leibniz had believed the monads to be without windows, the discovery of history in the 18th century opened the understanding to a view of the formation of individuals in community with others. This means that we wouldn't exist if we were limited to ourselves. Kant himself recognized history in his later years. Goethe, therefore, was justified in renouncing critical or romantic narcissism and to turn from Kant's intellectual intuition to an intuitive reasoning which is experienced as proceeding from the whole to the parts. Just this was the crucial point in his conversation with Schiller. Goethe considered the 'idea' to be Nature itself in one of her many forms, or as God's creative power in Nature. This idea was of course not to be grasped like Kant's subjective ideas. To become aware of it meant to *overcome oneself* "with reason, intellect, imagination, belief, feeling, mania, and if nothing else helps, with foolishness" (HA XIII 31). Goethe was aware "that we start theorizing with any attentive glance at the world (daß wir schon mit jedem aufmerksamen Blick in die Welt theoretisieren)" (HA XIII 317). This, however, is not our unsurmountable fate or subjective restriction but we must ask ourselves again and again: "Is it the object or is it you who expresses himself here? (Ist es der Gegenstand oder bist du es, der sich hier ausspricht?)" (HA VIII 306). Elsewhere irony, fear, and awe (Ironie, Angst, Scheu) are added to the list of appropriate attitudes in order to become aware of the 'idea' in Nature by overcoming the subject's insularity.

This explains what in Goethe's letter to Wackenroder was meant by overcoming oneself instead of the object. Scientific knowledge demands acknowledgement of the object. In other words: We are supposed "to appropriate ourselves to Nature", not the other way around (sich ihr – der Natur – anzueignen: HA VIII 467). In this way Goethe believed that we could "by intuiting an ever-creating Nature make ourselves worthy of participating intellectually in her productions (daß wir uns, durch das Anschauen einer immer schaffenden Natur, zur geistigen Teilnahme an ihren Produktionen würdig machten)" (HA XIII 30f.). This is Goethe's paradigm for the study of Nature: By intuition of an ever creative Nature we must qualify ourselves to participate in her productivity. Have we done so in modern biotechnology? Have we qualified for participation in Nature's productivity by overcoming ourselves in this respect? With these questions we are prepared for Faust's, or rather Wagner's, Homunculus.

Wagner, Faust's eager assistant from the first part of the drama, has finally managed to become his successor as the director of the institute. Mephistopheles happens to enter his biochemical laboratory at a crucial moment. "A man is in the making (Es wird ein Mensch gemacht)", Wagner whispers to him, and when Mephisto inquires which loving couple had been hidden in his experimental set-up, Wagner replies that the traditional kind of reproduction was to be left to the animals while "A man with his great gifts must henceforth win / a higher, even higher origin (so muß der Mensch mit seinen großen Gaben / Doch künftig höhern, höhern Ursprung haben)" (Vs. 6835, 6846f.)

Did Wagner overcome himself in order to participate in the productivity of Nature? His main concern was to appropriate the mysteries of Nature for the promotion of Man and to release the mind from the body, giving Man a 'higher origin' than a loving couple. This was certainly not the kind of science Goethe would approve of. What then does modern molecular biology look like if considered according to Goethe's crucial question: Is it the object or is it you who expresses himself here?

Man as the object

It is easy to give examples of how science and technology would have been better if we had tried to participate in the productivity of Nature. For instance:

- Nature avoids energy intensive high temperature processes in organic growth. Photosynthesis is a much better energy system than nuclear fission or burning fossil fuels.
- Nature is very careful with toxic substances such as Chlorine.
- Nature has developed materials like wood which are most durable in use and can easily be degraded afterwards. CFC or DDT are not of this kind.

Nature's 'soft biotechnology' is certainly a better paradigm for the industrial economy than traditional chemistry. It is interesting to consider how differently molecular biology might have developed if what had proved successful in the history of Nature had been taken as a paradigm.

At the present time, however, the crucial question with respect to cultural limits in science and technology is Man himself. Since we have emerged from the history of Nature, Man himself may become an object of knowledge in Goethe's sense, qualifying as participant in the productivity of Nature. In fact, 'Know thyself' has been a directing principle in human life for millennia. There is a project in current biotechnology which claims to provide just this, self-knowledge. I am referring to the Human Genome Project which is pursued in the USA and in Europe. Robert Shapiro has written a book on the project and its motivation. I am told that he is one of the leading participants who really knows what he is talking about. Reading his book, "The Human Blueprint" (1991), answers the question of whether the project goes beyond the limits of culture in Goethe's sense. The answer is clearly, Yes. I want to mention four basic ideas which are at the forefront of the Human Genome Project and which do not qualify as being compatible with participation in the productivity of Nature.

(1) While Wagner's motivation was to provide Man with a "higher origin" than carnal love, Shapiro praises molecular biology for the discovery that Man is a text, almost pure spirit, or a blueprint in DNA-letters. This text, Shapiro remarks, happens to become conscious of itself in a body which wasn't its choice and in a world it has not created itself (xvi). But by now we know the blueprint and can – as we do with the external world – take over responsibility for the maintenance, improvement, and reshaping of our body (215). This is comparable to repairing or remodelling an automobile (xix). We want to find out how our body works, what makes us ill and why we must die (xviii). Our goal must be to improve humanity by conscious evolution (374) and the first step to doing so is the Human Genome Project.

This sounds like emancipation. Bodily, we have emerged from natural history and the idea seems to be that now the time has come to overcome our personal dependence on matter. I doubt whether this is true, because I am aware of myself as *being* this body as long as I am living here and not *having* it. To consider Man as a text, or an intellectual structure, and the rest of the world as a bag of resources, is a Cartesian idea which has led us into the environmental crisis. Instead of renouncing the dualistic error, the geneticist's idea now seems to be that after having destroyed so much in Nature we ought to sneak away from it as far as possible. Certainly this does not qualify us as participating in the productivity of Nature.

The Cartesian outlook is made explicit by some technical ideas suggested by Shapiro for improving our body and preserving our mind, for instance:

- *immortality*, the idea that by means of the genetic disc – the minimum embodiment of one's personality – reincarnation could be achieved like the repeated performance of a play in a theatre (362f.);

- *the individual's lasting renown*, by means of that same disc so that future generations even after ten thousand years can find out who and how their ancestors were – while e.g. even on the moon only the name of a US president has been engraved, which contains much less information than the disc (358);
- *the genetic supermarket* for parents to compile the properties of their children (370f.). The properties on the shelves which Shapiro mentions as genetically conditioned are, for instance, height, colour of skin, complexion, bodily and mental strength, hair, intelligence, alcoholism, left-handedness, and colour blindness.

Shapiro also mentions that already the genetic text is much more reliable for cutting down on crime than fingerprints and that Oedipus could have known about his parents before he killed his father and married his mother. Even the Unknown Soldier, he adds, must no longer remain unidentified (302, 278, 292). If some of this sounds ridiculous, it is yet characteristic of the attitude.

Apart from Shapiro's Cartesianism, the other three ideas I want to pinpoint with respect to the limits of culture in Goethe's sense are, the elimination of disease, safety in basic decisions about one's life and the social relevance of 'the disc' in a competitive society.

(2) Genetic therapy of diseases is generally given as the main goal of the Human Genome Project. Shapiro, however, doesn't restrict himself to the usual examples like sickle-cell anaemia, cystic fibrosis, Huntington's disease and cancer, but goes further to include high blood pressure and diabetes. He also is frank enough to admit that from the inner logic of this approach it is not reasonable to stop at somatic therapy, which is certainly true. The goal then really is the improvement of the human body by the elimination of disease. In fact nobody likes to be ill. Yet any disease is one way of settling a conflict, as psychosomatic studies have shown. To be sure, a disease is only the second best way to do so after the healthy attempt to come to terms with adversity. But what would happen if this second choice were no longer available? Viktor von Weizsäcker – well-known as the founder of psychosomatic medicine – has suggested that in this case moral wars would replace disease and that these wars would cause far more suffering than disease hitherto (VII 383). This may be right or wrong but in any case there are good reasons to believe that diseases as well as their treatment – but not their elimination – are part of Nature's productivity and in so far a part of culture.

(3) The Genome Project is bound to result in very personal implications. Shapiro gives the example of a young man who orders his personal DNA disc at a pharmacy in the year 2020 and picks it up an hour later. The disc will tell him what people in earlier times expected to find out from astrology, namely the propensity for all the good and bad that can happen in one's life, e.g. special talents like those for mathematics or music, as well as risks like those of growing fat and getting diabetes or other diseases. In fact, genetic

knowledge combines the astrologist with the insurance salesman, except that this man even allows the possibility of avoiding misfortune, not just paying for the damage. The combination of astrology and advanced insurance offers quite a chance to play it safe in life, and this is appreciated by the young man. Now – in Shapiro's book (279) – Millennia comes in, a nice girl whom he likes. But before falling in love with her, he of course wants to see her disc and find out whether their common risks are below the individual ones. She, however, wants to be loved for her own sake and not because of her DNA. Also she wants to follow her feelings and not the scientific interpretation of her genetic structure. In this conflict Shapiro clearly decides in favour of the young man (310). As opposed to him, I think that consciousness about ones feelings is a gift of Nature which can provide an appropriate degree of safety in an uncontrolled life. The young man wants to optimize his life but Nature never optimizes – this is her way to be prepared for unexpected events. The geneticist, therefore, again does not seem to participate in Nature's productivity.

(4) From natural history Man has emerged as a social being. One isn't oneself just for oneself. The social institutions which determine the individual must, therefore, be taken into account when remodelling of the individual is considered. In our times one must imagine what will happen if individuals without certain innate properties are exposed to competition – be it in the economy, in science, or in politics – but could achieve advantages in competition by changing their properties for instance by drugs, or if their parents could do so from the genetic supermarket. History means that given conflicts tend to be settled to the advantage of some and to the disadvantage of others. If genetic opportunities to 'improve' human nature with respect to certain contests are introduced at a given stage of history this would be as if the rules of play were changed when the game is already going on. This may be done, but one ought to consider the winners and the losers. It seems to me that the social and political aspects of human nature are missed out here, so that again I do not find the Human Genome geneticists on the side of culturally participating in Nature's (social) productivity.

My argument does not imply that the chemistry of life is generally the wrong approach or not worth studying. Goethe himself was in favour of it and even considered inventing plants. But my analysis clearly shows that science so to speak is not as scientific as its results. The results, for instance in human genetics, are scientific answers to questions which we call scientific as well. Are the questions, however, as scientific as the answers so that they can equally be considered true or false? While the correctness of the results can be proven scientifically, this is not the case with the questions. Whether the Human Genome Project is on the right track in answering the particular questions I referred to may be called a matter of appropriateness. From Goethe's paradigm of studying Nature, the project turns out not to be appropriate, i.e. to give right answers to inappropriate questions. I believe this to be true, but neither Goethe nor Rudolf Steiner nor any other great man or woman can settle this matter for us. The questions we are faced with must be answered by ourselves out of our own responsibility. The main outcome of my analysis is that this discourse ought to take place and to be centred on the appropriateness of the questions to which DNA research gives the answers.

Whether a particular kind of knowledge is appropriate, depends on our position in Nature. In my physiocentric view, the chance to bring about culture as a specifically human contribution to the evolution of Nature depends on identifying ourselves not only with our bodies but with the sensual world itself. Nature can flourish in Man if we participate in her productivity. It seems to me that biotechnology in general could fit into Nature's productivity if it were embedded not only in a chemical but in a so to speak really biological biology (cf. Jax 1996). I would like this conference to provide some guidelines for just such a biological future for DNA.

References

Goethe, J.W. von (1981) *Goethes Werke*, E. Trunz (ed.), Hamburger Ausgabe in 14 Bänden, Beck, München.

(Apart from "Faust", quotations are given in my translation. "Faust" is quoted in G.M. Priest's translation.)

Goethe, J.W. von (1976) *Goethes Briefe*, K.R. Mandelkow (ed.), Hamburger Ausgabe in vier Bänden, 2. Aufl., Beck, München. Quotations are given in my translation.

Jax, K. (1996) Über die Leblosigkeit ökologischer Systeme, Zur Rolle des individuellen Organismus in der Ökologie, in H.W. Ingensiep and R. Hoppe-Sailer (eds) *NatürStucke – Zur Kulturgeschichte der Natur.* Ostfildern (edition tertium), pp. 209-230.

Meyer-Abich, K.M. (1993) *Revolution for Nature: From the Environment to the Connatural World*, The White Horse Press/Univ. of North Texas Press Cambridge, UK/Denton, pp. 145.

Meyer-Abich, K.M. (1997) *Praktische Naturphilosophie – Erinnerung an einen vergessenen Traum*, C.H. Beck, München.

Shapiro, R. (1991) *The Human Blueprint – The Race to Unlock the Secret of our Genetic Script*, St. Martin's Press, New York, pp. 412.

Weizsäcker, V. von (1987) Meines Lebens hauptsächliches Bemühen, in P. Achilles (ed.), *Allgemeine Medizin, Grundfragen medizinischer Anthropologie, Gesammelte Schriften Vol. VII*, Suhrkamp, Frankfurt/M. pp. 372-392.

Acknowledgement

I am grateful to Pat Cheney in England who converted my German English into her's most considerately.

3 The cultural powers of the gene – identity, destiny and the social meaning of heredity

SUSAN LINDEE
Historian
Department of the History and Sociology of Science, University of Pennsylvania
3440 Market Street, Suite 500, Philadelphia, PA 19104-3325, USA

This paper was communicated by Carla Keirns, Department of the History and Sociology of Science, University of Pennsylvania.

Abstract
The contemporary gene seems, in popular sources, to determine fate, identity and social place. In this paper, I explore how images of heredity and DNA facilitate institutional goals – for employers, schools and county – and how they define individuals as simply DNA writ large. I suggest that DNA functions as the contemporary equivalent of the Christian soul – containing the essence of the individual, and conferring genetic "immortality".

We recently attempted to search a widely used computer data base for references in the popular press to both "genetics" and "destiny." Nexis came back and told use that the search parameters were too inclusive – that they would produce more than a thousand references, and that we should try to focus my search more sharply.

Indeed, genetics and destiny are so tightly linked in contemporary culture that to mention one is almost automatically to invoke the other. They are linked in the world of biomedical science – where the presence of the BRCA1 gene seems to dictate future medical problems.[1] They are linked in the political world – where the authors of the The Bell Curve confidently argued that genes determine intelligence, intelligence determines success, and therefore genes determine success.[2] And genetics and destiny are linked in the everyday social world of individual decision-making and institutional controls.[3]

We explore here this linkage between genetics and destiny, considering the ways that both past and future seem to be written in the genes, and the ways that DNA functions as the contemporary, scientific equivalent of the human soul. DNA's special relationship to time has played an important role in its cultural meaning. In both popular and scientific sources, DNA appears as a molecular text that operates outside of time, preserving ancient human history in a permanent archive, and even revealing, with proper analysis, the human future, both collective and individual.

In some cases, DNA seems to bear witness to the remote human past, telling us stories about early human population shifts – "the demographic nature of the spread of farming technology in Europe around 10,000 years ago"[4] – to quote from a description of the Human Genome Diversity Project.[5] The ethical questions raised by this project have received a great deal of attention in various institutional settings, including the United Nations, human rights groups, scientific organizations and organizations of indigenous peoples.[6] But virtually all sides in these debates about the genome diversity project have accepted the idea that the genome is somehow sacred, that it occupies a special place in the cartography of the body – as one Native American protesting the project put it, DNA is not just "protein actions to be interpreted. For us," she said, "genes are sacred."[7]

They are also, of course, commodities in an international network of commercial and scientific exchange: The sacralized gene and the profitable gene co-exist in the same frameworks. It has become theoretically possible to sell one's (genetic) soul – particularly if the genes contained therein can be used to develop lucrative diagnostic technologies to predict future disease – for the genome where the human past is written also seems to contain stories about the human future.

In 1969 the American geneticist Victor McKusick announced that "genetic factors are involved in all diseases." He characterized the mutant gene as an "etiological agent comparable to a bacterium or virus." Almost thirty years later, McKusick's proposal has indeed become the conventional wisdom. Introductory medical textbooks in the 1990's feature charts showing disease ranked by their degree of heritability, Huntington Disease at one end, tuberculosis at the other. Promoters of molecular genetics promise a "fundamental transformation" in clinical medicine, comparable, they say, to "vaccines, antibiotics, antisepsis and anesthesia." And in one much-repeated narrative of public health, genetic disease grows in importance as other sources of disease recede. To quote one textbook, genetic disease has already become central to clinical practice in first World countries, and "the same will be true in the developing world once high mortality rates due to infection and malnutrition come under control."[8]

In popular and scientific sources, in textbooks, clinical practice, and legal policy, DNA has acquired immortality, canonical status as a sacred text, and predictive abilities. All disease, all success, all social disorder can be tracked, in these narratives, to a molecular structure. A particular terrain of the cell – historically crucial to several high-profile scientific and industrial fields in the late twentieth century – has emerged as the transcendent leader on the shop floor, and as an independent agent in human history.

What is this crucial entity? In one sense, the gene is a biological structure, the unit of heredity, a section of DNA that specifies the composition of a protein. But it has also become a cultural icon, a symbol, and a magical force. The biological gene has a broader meaning that is independent of its precise biological properties. Both a scientific concept and a powerful social symbol, the gene has many powers.

The limitations of these readings should be readily apparent to a scientific audience. DNA tells us some stories that are almost always true; others that are probabilistically true; still others that are true only under very specialized conditions. DNA also tells us stories that are not reliable at all. As a holy text, it has many of the failings of the Bible: Ambiguity, contradiction, obscurity, repetition.

But these failings and ambiguities disappear in many popular and scientific sources. Here, DNA exists in inanimate objects: A full-color advertisement boasts that a new BMW sedan has a "genetic advantage," a "heritage" that comes from its "genealogy;"[9] other ads announce that a Sterling's remarkable handling is "in its genes;"[10] that a Suburu is a "genetic superstar;" and that a Toyota has "a great set of genes."[11] The Infinity – another car – has DNA that defines its authenticity: "While some luxury sedans just look like their elders, ours have the same DNA."[12] Genes mark the quality of other products as well. A Nike sneaker "has inherited its own set of strength... resilience... stability, and a true intelligent fit."[13] A Bijan perfume called DNA is advertised as "a family value," inspired by the "power of heredity." It is "the stuff of life."[14] A blue jeans ad exploits the obvious pun: "Thanks for the genes, Dad."[15] An article on the leadership changes in the New Yorker asks: "Can you change a magazine's DNA?... A magazine's underlying character remains – unchanged and enduring, a DNA-like set of fingerprints – Tina Brown has much to reckon with, starting with 67 years of DNA."[16]

DNA is also commonly the focus of humor: A cartoonist lists genetically linked traits: "excessive use of hair spray, bottomless appetite for country-western music, right wing politics."[17] A barroom cartoon by Nick Downes portrays "Dead-Eye Dan, known far and wide for his fast gun, mean temper and extra Y chromosome."[18] Bad genes have become a facetious metaphor to describe national aggression. A magazine article on the "new Germany" describes the nation as "a child of doubtful lineage adopted as an infant into a loving family; the child has been good, obedient, and industrious, but friends and neighbors are worried that evil genes may still lurk beneath a well mannered surface."[19] A journalist describing violence at English soccer games blames it on the "genetic drive to wage war against the outlander."[20]

News reporters and talk show hosts refer to "bad seeds,"[21] "criminal genes,"[22] and "alcohol genes"[23]; CBS talk show host Oprah Winfrey asks a twin on her show whether her sister's "being bad" is "in her blood?"[24] In the movie JFK one character tells another "You're as crazy as your mama – Goes to show it's in the genes."[25] To one journalist, genes contain the equivalent of original sin, an evil that is "embedded in the coils of chromosomes that our parents pass to us at conception."[26]

News reports on diet control often begin from the assumption that genetics is the underlying reality that determines obesity; "Where Fat is a Problem, Heredity is the Answer," reads a newspaper headline.[27] Another journalist writes: "Smoking has to do with genetics, and the degree to which we are all prisoners of our genes... You're destined to be trapped by certain aspects of your personality. The best you can do is put a leash on them."[28] And in a "Dear Abby" column, a lifelong smoker announced that he had no intention of stopping. His reasoning? "Heredity plays a major role in how long we live – not diet and exercise, jogging and aerobics, or any of the other foolishness that health freaks advocate."[29]

Self help books devoted to coping with addiction speculate on why some people are affected by their circumstances more than others. The answer: biological predisposition. Addicts, those who shop, eat, love or drink too much are "victims of a disease process" and the disease – the tendency towards compulsive behavior – is biologically transmitted by their families. "Obsessive Compulsive Disorder is partially genetically transmitted. ... Most researchers agree OCD will develop only if an individual is genetically predisposed to it."[30] Similarly, in a guide written for the families of gamblers, we learn that some people have a "biological temperament" that makes them especially susceptible to addiction.[31]

The iconic gene can appear in unlikely places. A children's toy catalogue features a "truly amazing" specimen of DNA embedded in a "museum-quality lucite block."[32] A new biography of James Joyce contains a diagram of Joyce's "genetic make-up."[33] A critic, reviewing a play about the persistence of racism, says "it's as if [racism] has a DNA of its own."[34] A story about a promoter of rap music describes his style: "Charlie Stettler has no embarrassment gene."[35] And former New York City comptroller explains why she was attracted to social causes: "It's in my genes."[36]

At these stories suggest, in the 1990's, "gene talk" has entered the vernacular as a subject for drama, a source of humor, and an explanation of human behavior.[37] In supermarket tabloids and soap operas, in television programming, in women's magazines and parenting advice books, genes appear to explain obesity, criminality, shyness, directional ability, intelligence, political leanings, and even preferred styles of dressing. There are selfish genes, pleasure-seeking genes, violence genes, celebrity genes, gay genes, couch potato genes and depression genes. There are genes for genius, genes for saving, and even genes for sinning. These popular images convey a striking picture of the gene as powerful, deterministic, and central to an understanding of both everyday behavior and the "secret of life."[38]

What role have scientists played in this popular imagery? By valorizing this molecule – emphasizing its uniqueness – scientists have been able to marshal public support for scientific research to study it. This is a routine phenomenon in the history of science: The architects of emerging scientific fields commonly link their subjects of study to broader cultural themes, in an effort to provoke public interest and support. Thus geologists in late nineteenth-century America tied stratigraphy to manifest destiny, constructing their knowledge of rocks as necessary to the development of the American West; And chemists in eighteenth-century Britain promoted the relevance of chemical theory to the practical needs of a colonial empire. So too genomics scientists in the late twentieth century have successfully operated within a cultural terrain in which the control of disease, mental illness, crime and other social problems has acquired all the urgency once focused on the control of the West, or the control of the empire. The genome has become a profitable "new frontier," a site for international competition, and a potential guide to the elimination of a broad range of human suffering.

Many dramatic claims about DNA have been made by leading genomics scientists who have called the genome a "Delphic Oracle," a "time machine," a "trip into the future," a "medical crystal ball." It is a "Bible," the "Book of Man," and the "Holy Grail." Former director of the US Human Genome Project and Nobelist James Watson has proclaimed that DNA is "what makes us human"[39] and that "our fate is in our genes."[40] Some scientists have even claimed that our genes – or our ability to control them – can lead us to a land free of illness, crime, uncertainty or psychic distress. One geneticist promises that with the help of gene therapy "present methods of treating depression will seem as crude as former pneumonia treatments seem now."[41] And a biologist and science editor, describing acts of violence, has suggested that "when we can accurately predict future behavior [with genetic analysis], we may be able to prevent" such behavior.[42] Meanwhile a molecular geneticist of our acquaintance, in a casual conversation, happily predicted a future system of infant assessment: All newborns, he said, will one day be given a printout of their own genome, along with some frank advice about talents, deficits and ideal career choices.

"Is DNA God?" asks a skeptical medical student in an essay in a medical journal: "Given [its] essential roles in the origin, evolution and maintenance of life, it is tempting to wonder if this twisted sugar string... is, in fact, God."[43] Such language conveys an image of a molecular structure as more than a powerful biological entity: it also appears as a sacred text that can explain the natural and moral order – and here, too, molecular biologists have played an important role. Indeed, in at least one commercial plan the connection between the scientific promotion of DNA and the popular understanding of its miraculous powers is quite explicit: Kary Mullis' Stargene, a biotechnology firm, produces "celebrity DNA" cards for sale to teenagers. These cards contain bits of DNA from famous persons, amplified by PCR – the technique for which Mullis won the Nobel Prize in 1993 – and they function as molecular relics.[44]

Mullis has explained that the purpose of such cards is to educate people about DNA and about evolution. Samples from various primates, he has suggested, can be used to help school children understand genetic change through time.[45] But of equal importance, the cards could bring molecular biology to the people: "People could use the cards as totems or relics, but they could also learn about genes by comparing different stars' sequences."[46]

Mullis' DNA cards can be understood as a form of contagious magic, the mystical construct that underlay the widespread distribution of pieces of the True Cross and other Christian relics in the fourth and fifth centuries. In contagious magic, any object that comes in contact with a revered person, or any part of that person's body such as hair or bone, is believed to be equivalent to the person's whole self, no matter how small, or how distant in time. A piece of bone, a single hair, or a bit of cloth or wood from an object once touched by the person can, in the words of the New Catholic Encyclopedia, "carry the power or saintliness" of the person "and make him or her 'present' once again."[47] Such objects are commonly called relics, and they played an important role in early Christianity. At the height of the "cult of relics" fashionable noblewomen wore around their necks amulets containing splinters of the True Cross – the cross on which Christ died. Eventually wood from the True Cross "filled the world" though "miraculously the original cross remained whole and undiminished in Jerusalem."[48] The rage for relics had the advantage of bringing the saints directly to the people, and the remains of saints became a symbolic exchange commodity that fostered the spread of Christianity at a pivotal time in Church history.[49]

Like the True Cross in the early Christian period, the bits of celebrity DNA produced by Kary Mullis and his company could "fill the world" without becoming depleted. Like early church leaders, Mullis is interested in spreading the faith by bringing celebrity DNA to the people. Molecular relics promise to make the revered person "present" for the follower. And like relics in the fourth century, DNA cards will educate their owners, enrolling them in the molecular paradigm. Mullis is explicit about this agenda: he intends the DNA cards to be a form of popular promotion of molecular genetics.

Molecular relics have also appeared as the essence of immortality in the investigation of Lincoln's DNA. It is a case in which DNA seems to contain crucial historical details about an admired person. When an American museum began its study of the DNA contained in preserved bits of Lincoln's hair, bone and blood, scholars theorized that Lincoln might have suffered from Marfan syndrome, a rare genetic condition characterized by weaknesses in bones and joints, eyes, and heart. Anecdotal evidence links Marfan to high intelligence, and Marfan patients are often tall, with long limbs and long fingers – as was Lincoln.

The primary risk in the condition is that the aorta will burst – many Marfan victims die relatively young as a consequence of heart problems. And so the historical debate about Lincoln as a victim of Marfan syndrome has explored whether the disease could have taken his life at any time even if John Wilkes Booth had failed to assassinate him in April of 1865. "Was the slain president doomed by a disease?" asked a newspaper headline. The "genes define the essence of the person," noted one journalist, covering the debate over Lincoln's DNA: "some scientists suggest that genetic evidence might also one day show whether Lincoln suffered from chronic depression, as several biographers suspect, or from other conditions that affected his decision-making."[50]

In this narrative, President Abraham Lincoln – the entire social, historical, cultural and biological actor – can be retrieved from relic-like body parts stored in museums in

Washington, DC. His DNA seems to "make present" the historical figure in all his complexity. Molecular analysis of DNA can reveal the structure of his intelligence and his emotional state, and even his decision-making style. And unlike Lincoln's own writings, his speeches, his correspondence or the correspondence of those who knew and observed him in action – unlike these archival documents chronicling his actions and his words – DNA can tell us what his true fate would have been had he not been killed by an assassin. Indeed, Lincoln's DNA, taken from his remains, is like an immortal or eternal text that can be deciphered by contemporary molecular biologists. Identity, self, future and past are all apparently contained in the bits of molecular material that can be retrieved from bodily remains, fingernails, shattered bone, or blood scattered at a crime scene.

Those participating in the commercialization of DNA in the biotechnology industry may have a particularly strong stake in a construction of the gene as the equivalent of the soul – the source of identity and destiny. A new biotechnology firm in Houston, Texas, for example, has announced a commercial service to freeze tissue from aborted fetuses – for a $350 fee – promising clients that the aborted infant (apparently fully contained within the frozen tissue, in the DNA within the cells) can be reconstituted at a later time, through hypothetical technologies – a fetus with the same DNA, therefore, but born some time later, apparently qualifying as the same fetus.[51] Are two people who have the same DNA the same person? Not usually – certainly not in the case of identical twins or triplets. But the idea that they are somehow the same person persists, even in promotions of DNA by scientists.

Thus anthropologist and popularizer of socio-biology Melvin Konner suggests that twins separated at birth, when reunited as adults, experience a "strange boundary-blurring union." Who am I? is one of the most basic human questions, Konner notes. Meeting another human being who is genetically identical is therefore, he says, a jarring experience, a challenge to self-actualization.[52] Richard Dawkins, in his popular book The Selfish Gene, called human beings "survival machines – robot vehicles that are blindly programmed to preserve the selfish molecules known as genes."[53] Dawkins may seem wildly materialist and anti-religious, but his extreme reductionism, in which the DNA appears as immortal and the individual body as ultimately irrelevant, is, in many ways, a theological narrative, resembling the belief that the things of this world (the body) do not matter, but the soul (DNA) lasts forever. Like the frozen fetal cells and the concerns about boundaries and genetic individuality, the idea that genes "live forever" derives more from cultural context than from any necessary conclusions about cellular processes.

In her study of theological debates about the soul in the twelfth and thirteenth centuries, historian Carolyn Bynum explores how the issue of personal continuity – the survival of the self or soul – has long focused on actual physical body parts. Bynum herself connects these theological debates to twentieth century philosophical explorations of brain transplants and beamed up characters in Star Trek. When we think about identity, she points out, we think about bodily materials, and in many important ways the body is the person – containing memory, history, and future. Early Christian thinkers, as they explored "how identity lasts through corruption and reassemblage" of the body, debated whether discarded fingernails would be reunited with their rightful owners at the end of the world; they wondered what age people would be when they rose from the dead, and whether those who had lost a leg, arm, or eye, would regain these pieces of themselves on the day of Last Judgment. It is the gruesome physicality of these debates – their obsessive concern with actual biological materials, and their assumption that soul or self was contained within biological materials – that make them relevant to our discussion today.[54]

For the modern cultural concept of genetic essentialism draws much of its power from such theological roots – the idea of genetic destiny and immortality in the genes resonates with these early Christian debates about lost fingernails: Like the Christian soul, DNA is an invisible but material entity, an "extract of the body" that "has permanence leading to immortality."[55] And like the Christian soul, DNA seems relevant to concerns about morality, personhood, and social place.

The gene has become a way to talk about the boundaries of self, the nature of immortality, and the sacred meaning of life in ways that parallel theological narratives. The similarities in the powers of DNA and those of the Christian soul are more than linguistic or metaphorical. DNA has taken on the social and cultural functions of the soul.

How has this sophisticated scientific entity (which can be sequenced, PCRd, flow sorted) acquired mystical properties? I would suggest that DNA has been sacralized not by the vulgar masses – not by journalists or script writers or advertising executives – but by molecular geneticists and other scientists, who effectively repackaged the most powerful and compelling images of one belief system to meet the needs of another: Those describing and promoting DNA have drawn on the most powerful images of Christianity to convey its importance.

DNA is thus the central player in several interrelated stories. It is the reason for human exceptionalism (the locus for human evolution) and the source of human problems (social disorder deriving from genetic decline). Past and future, identity and destiny, the soul and the self, all seem to be contained in the genome.

A recent exhibition by the conceptual artist Larry Miller featured printed documents offered to gallery visitors. This "Genetic Code Copyright" could be filled in by the visitor. The text read: "I_______being a natural born human being... do hereby forever copyright my unique genetic code, however it may be scientifically determined, described or otherwise empirically expressed, Sworn and declared by me, an original human, with fingerprint affixed herein." For $10, Miller would witness the completed forms for visitors.[56]

Miller's whimsy – posing as art – captures the central issues we have explored – the commercialization of the body and bodily materials, the sanctification of identity in the genome, and the resulting profound confusion over the nature of DNA, its meaning for science and its meaning for culture. Medieval theology and twentieth century capitalism have merged to produce a remarkable creature, a chimera, the modern genome – soul, commodity, and permanent historical archive. The process that built the monster is not technological but cultural; it is the accrual of layers of management and meaning through which we make science, and we make the world.

Notes

1 For a review of the status of scientific research on BRCA1, see L.A. Cannon-Allbright and M.H. Skolnick, (1996) "The genetics of familial breast cancer" Seminars in *Oncology* 23 (1 Supplement 2), pp. 1-5.

2 Herrnstein, R.J. and Murray, C. (1994) *The Bell Curve: Intelligence and class structure in American Life*, Free Press, New York.

3 See Nelkin, D. and Lindee, M.S. (1995) *The DNA Mystique*, Freeman and Company, New York; also, Nelkin and Tancredi, *Dangerous Diagnostics*.

4 Quoted in "The Case for the HGD Project: A scientific contribution to world culture" in Summary Document, Human Genome Diversity Project, at *http://www-leland.stanford.edu/group/morrinst/alghero.html*, printed on 6 September 1996.

5 Scientists involved in this project will use DNA collected from about 500 ethnic groups around the world, including the Yanomami of Venezuela and the Chukchi of Northern Siberia, to reach conclusions about the distant human past and about racial relationships. They hope to explain the Bantu expansion in Africa, the origins of Native Americans and the timing and number of their migrations across the Bering Strait, and the relationships between linguistic groups around the world. Leslie Roberts. 20 November 1992. "Genome Diversity Project: Anthropologists climb (gingerly) on board" *Science* 258: pp. 1300-1301. See also the Senate Committee on Government Affairs, US Congress, "Hearings on the Human Genome Diversity Project," April 26, 1993.

6 From "The Case for the HGD Project" op cit.

7 At a meeting in Paris of Unesco's International Bioethics Committee in September 1995, a Paiute Indian from America proclaimed that the Human Genome Diversity Project – the so-called "vampire project" – was "not a priority for indigenous peoples." She invoked the 1964 Helsinki Declaration – which states that "in research on man, the wellbeing of the subject takes precedence over science and society" – to urge that HGDP be terminated. The project, she said, would not meet pressing needs for basic human rights in indigenous people. Quoted in D. Butler (1995) Genetic diversity proposal fails to impress international ethics panel, *Nature* 377: pp. 373.

8 Victor McKusick, (1969) *Human Genetics* Englewood Cliffs, NJ: Prentice-Hall, 11.

9 BMW Advertisement, c. BMW of North America Inc., 1983

10 *Time*, November 5, 1990.

11 *New York Times*, May 1991.
12 New Yorker, October 5, 1992
13 *Vogue*, October 1991.
14 Mirabella, January 1993. The bottle pictured in some ads for Bijan's DNA has the amazing shape of a triple helix.
15 C. Klein, cited in A. Fausto-Sterling (1985), *Myths of Gender*, Basic Books, New York. pp. 7.
16 Diamond E. (1991), Can You Change A Magazine's DNA? *New York Magazine*, July 20, pp. 27.
17 Cartoon by R. Chet, *Health*, July/August, 1991, pp. 29
18 Downes, N. (1992) cartoon in *Science*, **256**, pp. 547.
19 Jackson, J.O. (1992) The New Germany Flexes Its Muscles, *Time*, April 13, pp. 34+.
20 Lehman-Haupt, C. (1992) Studying Soccer Violence by the Civilized British, *New York Times*, June 25.
21 Newman, M. (1991) Raising Children Right Isn't Always Enough, *New York Times*, December 22.
22 Fallows, F. (1985) Born To Rob? Why Criminals Do It, *The Washington Monthly*, December, pp. 37.
23 Nobbe, G. (1989) Alcoholic Genes, *Omni*, May, pp. 37.
24 "Oprah Winfrey Show," *CBS*, August 24, 1992, 4:00 p.m.
25 JFK. Warner Bros film, 1991
26 Franklin, D. (1989) What A Child is Given, *New York Times Magazine*, September 3, pp. 36.
27 See, for example, Irish Hall, Diet Pills Return as Long Term Medication Not Just Diet Aids, *New York Times*, October 14, 1992; and Kolata, G. (1990) Where Fat is a Problem Heredity is the Answer, *New York Times*, May 24.
28 Mansnerus, L. (1992) Smoking: Is it a Habit or is it Genetic, *New York Times Magazine*, October 4.
29 Dear Abby Column, Smoker has no plans to kick lifelong habit, *Delaware County Times*, April 7, 1993, p. 49.
30 Bar, L. (1992) *Getting control – Overcoming Your Obsessions and Compulsions*, Plume, New York, pp. 12-13.
31 Berman, L. and Sigel, M.E. (1992) *Behind the 8-Ball: A Guide for families of Gamblers*, Simon and Schuster, New York, pp. 43-44.
32 *Brainstorms Holiday Gift Catalogue*, 1995 Fall Edition, pp. 58.
33 Costello, P. and Joyce, J. (1993) *The Year's of Growth*, Pantheon, New York; Lehmann-Haupt, C. (1993) observed the focus on genetics in a review in the *New York Times*, April 8.
34 Rothstein, M. (1991) From Cartoons to a Play about Racists in the 60's, *New York Times*, August 14.
35 Stengel, R. (1992) The Swiss School of Rap, *New York Times*, October 18.
36 Quoted in Ley, F. (1994) Pride and Some Regrets, *New York Times*, January 1.
37 Howe, H. and Lynn, J. (1992) Gene talk in socio-biology, in Fuller, S. and Collier, J. (eds.) *Social Epistemology* **6**, No. 2, April-June pp. 109-164

38 "The Secret of Life" is the name of an 8 hour NOVA series, directed by Graham Chedd, and aired on public television, Channel 2, on September 26-30, 1993.

39 Jaroff, L. (1989) The Gene Hunt: Scientists launch a $3 billion project to map the chromosomes and decipher the complete understanding for making a human being *Time* 20 March, pp. 62-71.

40 Jaroff, L. (1989) The Gene Hunt, *Time*, March 20, pp. 62-67.

41 Wingerson, L. (1982) Searching for Depression Genes, *Discover*, February, pp. 60-64.

42 Koshland, D. (1992) Elephants, Monstrosities and the Law, *Science*, 255, February 4, pp. 777.

43 Henderson, G.S. (1988) Is DNA God? *The Pharos*, Journal of the Alpha Omega Alpha Honor Medical Society, Winter '88, pp. 2-6.

44 Kary Mullis *Omni* April 1992 69-92.

45 Weiss, R. (1992) Techy to Trendy, new products hum DNA's tune, *New York Times*, 8 September.

46 Weiss, R. op. cit.

47 "Relics" *New Catholic Encyclopedia*, 1967, pp. 276.

48 *New Catholic Encyclopedia* 1967, pp. 278. In the ninth century, a corporation was formed to discover, sell and transport relics throughout Europe. False relics proliferated, particularly in the wake of the Crusades in the thirteenth century. The commercial practices that built up around relics became a focus of Protestant satire. Thus John Calvin's (1509-1564) treatise on relics said the bits of wood from the "True Cross" were so numerous and heavy that not even three hundred men could have carried such a cross.

49 See Brown, P. (1981) The Cult of the Saints: Its rise and Function in Latin Christianity, University of Chicago Press, Chicago, pp. 89. Hinduism and Judaism have both rejected relics, possibly because in both traditions the bodies of the dead are interpreted as impure. But Buddhism and Roman Catholicism have both accepted and even endorsed relics.

50 Leary, W.E. (1991) Scientists seek Lincoln DNA to clone for a medical study, *New York Times*, 10 February.

51 The case is discussed in a newspaper column by University of Pennsylvania bioethicist Arthur Caplan that appeared on 17 July 1996 in the *Albany, N.Y. Times Union*, *Philadelphia Inquirer*, *St. Paul Pioneer Press* and other papers.

52 Konner, op. cit, pp 223-224

53 Dawkins, The Selfish Gene, op. cit., pp 24, 36

54 Bynum, C. (1991) Material continuity, personal survival and the resurrection of the body: A scholastic discussion in its medieval and modern contexts, in Bynam, C. (1991) Fragmentation and Redemption: Essays on Gender and the Human Body in Medieval Religion, Zone, New York, pp. 239-297.

55 These quotes are taken from Crawley's, A.E. (1909) The Idea of the Soul (a study dedicated to eugenicist Francis Galton), Adam and Charles Black, London, pp. 209, 211, 239.

56 The exhibit was "Genus Memorialis: Imitations of Immortality" at Horodner Romley Gallery, in Manhattan, February 1993. For a reproduction of the Genetic Code Copyright form, see Harper's, April 1993, 17.

4 The archetypal gene – the open history of a successful concept

ERNST PETER FISCHER
Biophysicist
Universität Konstanz
Postfach 5560, D-78434 Konstanz

Abstract
A review of the history of the gene and a look at the current usage of the term allow the observation that "gene" is a fuzzy entity with no clear-cut definition in sight. There is no number of genes in a genome. It is argued that fuzziness is a prerequisite for a successful scientific concept that has meaning beyond the limits of a given discipline (e.g. molecular biology), and it will be proposed that the endurance of the gene and the fact that science cannot do without it points to the fact that a gene cannot be understood as an empirical entity but rather as an archetypal one. Science seems to be full of archetypal concepts (atom, field, energy) and efforts are necessary to reveal their background.

Introduction
At the outset I would like to phrase an odd question: Does genetic engineering have anything at all to do with genes? When one applies the many methods collectively known as "genetic engineering" in order to recombine genetic material from biological cells in a test tube one deals first and foremost with a chemical substance – namely, that famous acid from the cell nucleus which even most lay people know by those three capital letters DNA. For a genetic engineer, the recombined molecule constitutes in a more or less clear-cut way what he considers to be a gene, but the question remains if this is the whole story? Does every biologist agree in viewing a gene as nothing more than a piece of DNA? Can any scientist say precisely which piece of DNA is meant? And is it just one piece, or are there many? Do we know where such a gene begins and where it ends? Or is there perhaps more to then concept of a gene than a molecule that can be found in a cell?

A fuzzy entity
The question as to what constitutes a gene will prompt quite different answers from different biologists, depending on their theoretical approach. In the field of genetics you

will find biochemists and evolutionary biologists, as well as classical and molecular geneticists, to name but a few. And they all think of something different when they hear or see the word "gene". To a classical geneticist, a gene is a unit of heredity transmitted according to Mendel's famous laws of inheritance. To a molecular biologist, it is a segment on a chromosome that codes for a unit of function (whether it be an RNA molecule or a polypeptide chain). For an evolutionary biologist, a gene is a cellular construction with sufficient endurance to serve as a basis for evolution. And for a biochemist, it is the information that a cell needs to synthesize a protein. And for a poet a gene is a gene is a gene.

With this brief excursion into the scientific variations on the theme, however, we have far from exhausted the subject. What we have not yet considered is the structural peculiarity of genes that was discovered in the early years of genetic engineering, when it was found, for example, that genes do not exist as molecular units, and that our cells contain genes instead in the form of multiple separate segments ("split genes"). Conceiving of genes in this way is an attempt to overcome their elusive nature, and as such it opens up many interesting possibilities (cf. the protean nature of antibodies, long a mystery to immunologists).

The question as to the precise nature of a gene has become an ever more urgent one, and one that today's observer of biology is compelled to face, not least because of the speculation that has raged for many years as to how many genes a human cell contains or needs – and still we await a satisfactory answer. What's the matter with scientists, can't they count? It must surely be possible to find out how many genes a person has? How many genes make up the human genome, i.e. the full set of genes in an individual? At least this answer should be coming out soon since scientists have defined the end of the Human Genome Project by knowing the complete sequence of all genes in a human cell.

At first this number was said to be about 50,000 genes. Later a figure of 100,000 was suggested. And then, when the first chromosomes (of yeasts and worms) came to have their molecular structure (DNA sequence) elucidated, some biochemists started talking in terms of half a million – how else were they to account for the fact that, in the DNA sequences published, they had found many more possibilities for unhindered processing of genetic information than could be explained by the number of proteins known to them at the time?

So we are faced with a confusing numbers game played out in front of biology students and slowly fostering the suspicion that the sought-after number does not even exist anyway, because no one can come up with a reliable definition of what constitutes a gene (Fig. 1). Indeed, I have the feeling that, once all the genome projects have been completed and everything we always wanted to know about sequences has been discovered, we are still not going to know how many of these things called gene are contained in a human cell. The molecular approach, after all, is only one of many possible ways of defining a gene. No discipline (I hope) will allow itself to be hindered from offering its own specific definition, even if the object of desire to which it was thought we had come so close with genetic engineering appears in the process to elude us.

Figure 1

Is a Gene a Structure in a Cell?

No Unit
Split gene and overlap
No Defined ends
Flanking and regulatory sequences
No Defined Location
Transposons and rearrangements
No Function
Pseudogenes and junk DNA

But does the gene actually elude us? In my opinion: no, not if we are prepared to entertain the notion that a gene is what could be called a "fuzzy entity" (Fig. 2). This would not appear to me to be any disadvantage. On the contrary, only by allowing "gene" to remain an open-ended concept that no one can precisely define and lay claim to for the purpose of his or her own discipline does it become accessible to many areas of science. Only in this way does it become the focus of attention continuing to hold a fascination and fostering the co-operation that helps to make research exciting. Only a fuzzy entity can appeal equally to the zoologist and the biomedical specialist. And those who are only amenable to persuasion at the molecular level need only consider the "fuzziness" that occurs through all the overlapping of functional sequences on which the specialist journals are constantly reporting.

The gene of course is not the only fuzzy entity in science (Fig. 3). After pausing to reflect for a moment the reader will have no problem finding his or her own examples of fuzziness, and will quickly realize that research is especially alive and kicking in those areas where it is not so easy to define precisely (and thus pigeon-hole) what is meant.

One has only to think of concepts such as "species" (from biology) or "self" (from immunology). And likewise "life" and "death" continue to defy attempts to settle on a clear definition. That the fuzziness of words is no disadvantage, however, is something we have to discover gradually, and the gene is a good place to start. It is the very vagueness of the term "gene" that allows it to say something about life which merits the attribute of truth. Indeed, truth and clarity are often mutually obstructive. The statement "Genes consist of DNA" is true chiefly because it remains unclear what genes are. If it were clear what genes are, then this statement – though not wrong – would be fairly trivial. Only a sensationalist would see in the statement anything more than a truism.

Figure 2

A Fuzzy Entity

Strong enough
to keep its identity
&
sufficient flexible
to be useful to more
than one group of scientists

Figure 3

Fuzzy Entities

Insulin
Physician vs. Chemist
Purity
Ecologist vs. Physicist
Information
Biologist vs. Engineer

A gene, then, is a fuzzy but successful idea which allows many different branches of science to understand how life is possible, how it reacts and develops. And it is precisely its fuzziness that constitutes the intellectual challenge and the fun.

A nice word

"Gene", however, appeals not only because of its fuzziness, but for another reason as well. Let us take a look, for example, at the origin of the term. The word "gene" was introduced at the beginning of this century by the Danish botanist, Wilhelm Johannsen, who suggested in his "Elemente der exakten Erblichkeitslehre" (1909) that the hereditary factors first speculated about by Gregor Mendel be given a name with a scientific ring to it. He proposed the monosyllabic word "Gen" (the original publication appeared in German). And the motivation, he admitted (and rightly so), was the simple aesthetic argument that it was an easy word to use.

Historically, it was extremely important that Johannsen chose a simple and easily pronounceable word, because when he was undertaking his research at the beginning of the century, his contemporaries first had to be convinced of the idea that concrete localized structures – namely genes – could be responsible for inheritance. Indeed, it took a few decades before embryologists, in particular, whose thoughts initially went in an entirely different direction, were finally persuaded. And today the paradigm shift is complete, as we all know. Today's geneticists and their colleagues are now so convinced that clearly identifiable structures called genes reside in all cells that they have – as mentioned earlier – started to count them. And they are certain, too, that eventually they will complete this task and be able to name a precise figure.

There are of course good reasons for assuming the concrete existence of "gene" in every cell. After all, the numerous proteins in a cell are a prerequisite for its survival, and the genetic information for each of these proteins must exist somewhere. In recent years, more and more of these structural genes have been isolated and sequenced – giving us genetic engineering and thus the possibility of modern genetics. In the process, scientists identified ever more yeast genes, mouse genes, goat genes, chimpanzee genes, and of course human genes. And clear though the information was that lay on the table before them, close inspection revealed it once more to be fuzzy and blurred, because the individual genes from the various organisms showed hardly any notable differences from one another. This higher form of agreement between species has long since been turned to advantage – for example, in transgenic animals. And it is now clear that a mouse cell can accept a yeast gene and a goat cell a human gene, just to name two of the many permutations.

The thing with two hands

A gene, then, cannot be just a clearly circumscribed piece of DNA arranged somewhere on a chromosome. It must be both closed and open. It must be delimited to be capable of recombination, and it has to be open if it is to function and be biologically active afterwards in a different setting. What unites the original environment (for example, a human chromosome) and the new setting (such as a yeast cell) is that all-embracing process we call evolution, what Charles Darwin described shortly before the discovery of genetic inheritance by Mendel. Genes are naturally made up of particles in the form of known cellular structures, but at the same time they are part of the continuum of information employed by nature in the development of life, in that process which we see as evolution and which brought humankind to the fore with such a flourish.

The question "What is a gene?" thus acquires an added dimension. It would seem that we can answer it in two ways. A gene is namely both a discrete structure and a continuous variable. This two-handed strategy is reminiscent of the response of twentieth century physicists to the question "What is the basic building-block of matter?" or "What is a unit of light?". Drawing on the principles of quantum theory, they say that these tiniest of units are both discrete particles and continuous waves. It proved difficult to gain acceptance for this view. In time, however, it emerged that only thanks to this approach was it possible to acquire that deeper understanding of the theory of matter which scientists were seeking and which continues to occupy the philosophical ground of science. And in the meantime it has been found that, while the truth may be stated about objects at the atomic or the subatomic level, this truth can only go so far – the things still retain their mystery.

What applies spatially to the elementary particles of physics, could – I suggest – be true temporally for the gene or genes. At any given time in the course of evolution – such as today, for example – a gene appears as a particular unit with a molecular structure (and given sequence). Seen over a period of time, however, it is a constantly changing phenomenon that joins with others to form a characteristic genome. The genome would then be a whole that does not consist of any parts at all, because the constituent genes only exist in the context, and by virtue, of this configuration. Here again is an analogy with physics, where it was realized that the atomic dimension is an embracing whole that does not consist of parts, since no elementary building-blocks of matter ever exist in isolation but only in conjunction.

The gene as cellular reality and open-ended potentiality, the gene as discrete unit and continuous information, the gene as concept with a sound base and fluctuating branches – however we look at it, the gene shows a dual character. And to this extent, it is and will remain a fuzzy object.

Let us now return to the question posed at the beginning of this essay, as to whether genetic engineering has anything to do with genes. In the light of the foregoing, it can be said that genetic engineering has only to do with DNA in the first instance, and anyone who applies the techniques of genetic engineering is attempting to find a

molecular answer to the mystery of life and to reduce the genes to DNA. In the process, however, they slip through his fingers, and disappear.
But this loss is in truth a gain, because thanks to this reductionist process one is ultimately left with a clear view of the interrelations outlined above. Only in the context of the possibilities opened up by genetic engineering do we grasp how the whole and its genetic parts fit together, and that they are essentially inseparable. Genes cannot be disconnected from life, however well genetic engineers learn to manipulate them.

The archetypal gene

Accepting the fuzzy nature of a gene is not possible without wondering why scientists stick to such a concept that they cannot define properly. The answer might be found in the observation that progress in science is sometimes possible with the help of entities that are not empirical but rather derived from internal sources. We can learn about that e.g. from Johannes Kepler, the great German astronomer of the 17th century (Fig. 4). In our time Kepler's ideas were picked up and given more details by the Austrian born physicist Wolfgang Pauli who was awarded the Nobelprize for physics in 1945 (Fig. 5). Pauli introduced the term "archetype" into this theoretical consideration to which he was introduced by the Swiss psychologist C.G. Jung (Fig. 6).

Figure 4

Johannes Kepler
in 1620:

We gain knowledge when we combine our perception with internal ideas and assess them to be in agreement.
What is inside will then light up in the soul.

Figure 5

Wolfgang Pauli
in 1950:

Theories are not made by logic conclusions but by bringing internal pictures into agreement with external objects. This is made possible by archetypes which govern what is inside as well as what is outside for man.

Figure 6

Archetype
Dictionary: The original pattern of which all things of the same type are representations. C.G. Jung: An inherited mode of thought that is derived from collective experience and present in the unconscious of any individual. W. Pauli: Archetypes belong to a reality without distinction between body and mind. They manifest themselves as mental images or physical laws.

I consider the concept of an archetypal origin of basic scientific concepts (Fig. 7) as an important idea that needs a serious effort in order to understand its relevance. In my opinion without such an archetypal origin – with understanding archetype along the lines of Pauli – we will not be able to understand basic historic facts such as the persistence of terms like atom or energy in spite of several complete changes of their meaning. There is no understanding of the history of science – and thus no theory of science – without finding what connects the inner world of ideas with the outer world of physical manifestations.

Figure 7

Hypothesis
Basic scientific concepts (e.g. atom, gene, field, wave, nucleus, energy, order) arise from archetypal images, which are at the origin of our thinking along with perception. In order to understand the history of science one has to learn more about both.

Thinking about archetypes is investigating the dark side of science, i.e. the side of science that cannot be looked at in the light of consciousness. We should not be afraid of it.

Reference

Fischer, E.P. (1995) *Die aufschimmernde Nachtseite der Wissenschaft*, Libelle Verlag, Lengwil

5 Back to the future – towards a spiritual attitude for managing DNA

JAAP VAN DER WAL
Anatomist-embryologist
Hogeschool Utrecht, Department of physiotherapy
Bolognalaan 101, NL-3584 CJ Utrecht

Abstract

The future of our society and culture will rather be destined by our mentality and our image of man and nature than by our knowledge and technology. Current concepts on heredity and DNA are the ultimate outcome of more than four centuries of Cartesian thinking about life, nature and man. Only the development of moral and mental points of view that are not directly in line with this out of date philosophical materialism and dualism might give us a directive for managing DNA instead of being ruled by DNA values. This contribution is a philosophical effort in order to save the reality of life and nature from the consequences of 'DNA-thinking' and sketches the outlines of an 'ethical biology'. It is stated that *gen-ethics* is far more important and shaping our future than genetics.

Introduction

This presentation deals with human consciousness, with mentality and in particular with (scientific) attitude. If we want to deal with future, we have to realize that not our technology determines how the future will look like, but that our mind, our image of man and nature is the central issue in this respect. It was like that in the recent past. It were not the inventions, machines or techniques that changed this planet and our society so rapidly and dramatically, but this all might also be considered as the appearance of evolving human consciousness and mind. In this way culture may rather be considered as a manifestation of human evolution than as the reverse (i.e. that modern human mind is the result or product of cultural evolution).

Nowadays science has overgrown its primary role as generator of facts about the 'world as it is'. More and more it represents an important – and for many people the only – generator of images of man and nature about the world 'as it really is'. Science has become the main source of knowledge to legitimate the 'world as it is'.

An issue of this contribution is that moral or ethical reflection on our attitude in dealing with DNA and genes should not be the last thing to do ('*after-science*') – something like a *value-added tax* originating from another domain than the scientific framework –, but that gaining awareness of ones scientific attitude is the first thing to do ('*pre-science*'). It will be shown how such a '*pre-scientific*' reflection may be performed and actualized, how it might play a role in our future with DNA and why scientists ought to feel responsibility for this reflection.

Back to the future?

A fruitful way to deal with expectations and with future might be a reflection on the past. This does not necessarily mean that on might predict the future by a simple extrapolation of the past. But a historical approach telling us how it came that we are and think as we do nowadays, might help us to reflect on the rightness or desirability of our ideas and concepts or at least become aware of implicit presumptions.

On the outer cover of your conference map you find the quotation of Francis Bacon upon the future of *homo manipulans* dating from 1624! "And we make by art [...] trees and flowers to come earlier or later than their seasons; and to come up and bear more speedily than by their natural course they do. We make them also by art greater much than their nature; and their fruit greater and sweeter and of differing taste, smell, color, and figure, from their nature. [...] By art likewise, we make animals greater or taller than their kind is. [...] Also we make them differ in color, shape, activity, many ways. We find means to make commixtures and copulations of different kinds; which have produced many new kinds, and these are not barren as the general opinion is. [...]. Neither do we this by chance, but we know beforehand of what matter and commixture what kind of those creatures will arise" (Holdrege, 1996).

In Holland, where this quote was the central item on a public debate in 1993 *Is there future in our DNA?* (Amons, 1994), many people were surprised that such a modern sounding citation comes from so far in the past. Does the quotation imply that everything that nowadays is very often presented as the superb future of biotechnology, is nothing more than very old wine (in methodological respect) in brand new bottles? Was Bacon clairvoyant? I think he simply deduced from the new way of thinking and perception (i.e. the new approach of man and nature that was coming up in his era of the seventeenth century) that events like 'prophesied' here would be the predictable outcome of that new way of thinking. Indeed it is inherent to Cartesian thinking that our world – and so future – is predictable: future comes out of the past.

The ghost of reductionism

In his recent book Daniel C. Dennett, the well-known philosopher of the human mind, proposes to differentiate between 'bad' and 'good' reductionism (Dennett 1995). He mentions 'bad' reductionism as the effort to reduce everything in living nature to the laws of physics and molecules. Dennett states that this reductionism is out of date and has been proven to be untrue and impossible. However he strongly defends 'good' reductionism. That is the effort to explain the world (man and nature) without the need of metaphysical entities or forces, without what he calls 'sky-hooks'. We do not need 'spirit' as secret or metaphysical principle behind things. Dennett calls us to 'become adult and grown-up and to accept finally that we are nothing but robots whose genes created us body and mind, like Richard Dawkins states' (Dawkins, 1976).

In my opinion there is no 'good' or 'bad' reductionism, Reductionism is methodologically good, but mentally bad, a real disaster for human awareness, for the human mind. It led, still leads and will lead to a corruption of reality, which nowadays incarnates in our society, economy, teaching, etc. and will lead to a fatal loss of bond with our roots. In order to elucidate this thesis I will search for the roots of reductionism. We will thereby be guided by the question that Davenport raised: 'Where is the wisdom we have lost in knowledge? Where is the knowledge we have lost in information?' (Davenport, 1978). I would like to extend this expression with: 'Where is the information we have lost in manipulation?'.

Two realities?

You are invited to attend with me the first encounter of medical students with the dead human body in the dissection room. It will be shown that there we may experience an *archetype* example of the real quintessence of reductionism. For more than twentyfive years it is my job but also my privilege to introduce medical students to the dissection room. In all those years I experienced in those introduction sessions the same typical reaction of the young students towards what happened to them.

I am used to confront the students first with the body of a dead person, intact as if he (or she) just had died. And there they stand, 'my' students, always surprised in any way, notwithstanding the kind of fore-knowledge they had or had not. There are emotions, feelings, to express and to be exchanged with each other, which I stimulate them to do. Some are anxious, some are horrified, others are moved and respectful. There are questions about Who and Why? For most of them a barrier of shyness has to be overcome to inspect and to touch the body. The dead person is recognizable to them from daily life ('look at those hands') but is also reminding them of human mortality, which also means their own mortality. The body as it lays there, expresses itself, has something to tell. The object (the dead body) and the observers (the students) as partners in dialogue. In methodological terms they are *participating* observers. Is this reality? No doubt! The experience is evident!

After that we usually turn our back to the body (literally) and direct our attention to a complete different body. The 'body by the anatomist', where the expert, the scientist has done his work. A lot is to be seen there: muscles, organs, nerves. And suddenly those attending student change! Their attitude changes, they turn into interested and eagerly curious people apparently without much hesitation or shyness. They start to grab things, turn them around, look behind them. They start to discuss with each other vividly. No private feelings and experiences, but questions like 'What is this? An aorta? How big, unbelievable?' Another body is at stake. The body of science, where their medical study largely will be based upon. But also true is that we deal with other students! They change position, they become *onlookers* (and with enthusiasm!), the body is object. Is this reality? No doubt! This body is also evident.

What happens here is that the position of the observer changes, therefore the relation to the 'object' is changed and so the observed. Reality is changing! That is a fact. The one who denies that and wants to defend that in both events the 'same' body is at stake, overlooks an essential threshold of difference in quality between the two bodies. He equalizes two qualities that are not similar or equal. In doing so he performs a so called *metabletic* jump! By this is meant that the observer changes reality without realizing he does so! (*Metabletics* = theory of changes). How easy this is to overlook! We are most of the time so unaware of our potency to change reality that we implicitly change from the one reality to the other without knowing or realizing. An example might explain this. *One* and *zero* are in the world within we live with our subject and senses, complete different and opposite qualities! That is evident! In mathematics however we approach from one (1) to zero (0) by minimalisation of the difference. We go to ½, to ½ of ½ i.e. ¼ and so on till the difference is that small that we make the smallest number equal to zero! Mathematically this is no problem at all. It is the essential of differential calculus. Mathematical laws stand it, but in the world of experienced reality we did something that cannot be done: we made a small but essential *metabletic* jump. In reality as we think it, we made two qualities equal that are not equal in the world as we experience it!

Back to the dissection room. My students undergo a *metabletion* in their relationship to the observed object: from partnership to onlookership. Some students still are in problems. They do not join their fellow students in the described latter round of the active enthusiasm of onlooking. Asked for it, they explain they hesitate because they remind the other, the former body. Have the others forgotten it? They, on their turn asked for that and having directed again their attention to the body they have left behind, say 'No, but that has nothing to do with this!'. Some students cry, others feel tears, even some of the enthusiast onlookers! – at least asked for it afterwards –. "Yes, it was like feeling I had to cry". But why? In all those years I searched for the nature of this emotion. It was such a common experience, it should mean something. One day a female student gave me the key. I checked it afterwards so many times that nowadays I am convinced that it is 'true'. She stated: 'I cried, because I did something I had to do – I want to become a doctor and have to study anatomy –, but at the same time something what I ought to do not. It feels like I left my youth behind, I lost something valuable but I had to. I will never be the same person as I was.' They leave a world behind! Has she become grown-up?

Separate worlds?

Do those two bodies, those two realities really have no relation? Are they separate worlds? And if there exists is a relation, what is the nature of that relationship? I always add a 'lesson' to the evaluation of this introduction session. 'Go and learn this body, this body of medical science with its anatomy, physiology, its evidence. But never forget that this body only exists in the very well defined relation of you as onlooker. It only exists on the surgeon table, in the X-ray picture, in the biochemical data and so on. But it will never be presented to you by the patient himself! He/she will present her living body which he/she is partner of. The patient will present as "I have pain, I am anxious, I am going to dy". All those qualities of the so called body of *given experienced reality* do not exist in the body of *thought abstracted reality.* In that body there is no pain, no fear, no mortality. Please, do not forget it. You might become a perfect doctor of THE body, but being a doctor for patients requires another consciousness, another attitude. And never say that the latter body is more "real" than the other. You experienced that today'. What we try to teach the students is that the relation between those two realities is hierarchical and not a matter of 'two of the same', of dualism. The one includes the other but not the other way around! With the realization of a medicine in which only the secondary reality of the 'human body' is the main topic, the other reality will succumb, will vanish. Returning to the philosophical telescope reveals that this event in the personal biography (*ontogeny*) of medical students and a certain anatomist reflects a homologous event in the cultural biography (*phylogeny*) of mankind in the modern so called western world. It represents an enormous *metabletic* turning point in our consciousness and relationship to the world, nature and ourselves. We have to point to the era of Bacon, the same Bacon that wondered us with his apparently prophetic words. In those era of the sixteenth and seventeenth century our attitude changed completely.

Literally that was exercised for all of us by genius people like *Vesalius*, the famous anatomist who was the first one to describe the human body with a modern scientific eye (1543). After Vesalius we definitively left behind the body of the Middle Ages! Many people think that the medieval scientist had no anatomy. The anatomy and physiology of the Middle Ages was not a abuse construct of theories, it was based upon the observation of the living body itself! I myself, being modern twentieth century anatomist, would be able to describe 'pre-vesalian' anatomy in complete evident and right terms, just basing myself upon the participating observation of the living human body. All kinds of experienceable facts are present there, which do not exist in the anatomy of the onlooking observation that we have developed after the sixteenth century.

The threshold that has been taken by Vesalius and that seems to separate us definitively from medieval 'pre-scientific' consciousness is clearly demonstrated comparing the portraits of two anatomists before and aftre the paradigmswitch. First the portrait of

Vesalius, this modern mind, as it appears on the front page of his book! This portrait radiates modern consciousness. Like the students on the dissection room: now we are the one at word, we will tell what the body, what the world really looks like! The awareness of the modern onlooker, with an eye and a mind to see a world full of detailed facts, with the mind to bring to light what it in there, in that body, behind the 'outside' of phenomena. Vesalius represents the modern awareness, *Gegenstand-bewusstsein* in philosophical terminology, awareness for details. Complete opposite consciousness appears to us from the illustrations by and the portrait of *Vigevano*, a medieval anatomist of two centuries earlier. He represents the awareness of *context*, unity, coherence. He cannot destruct, take apart in details. Many people think that the medieval anatomist were forbidden to dissect. That is not true. They could not. Why look in the dead for the understanding of life? Do you think that I could dissect my father? I as an anatomist? Maybe, but only if I think away the person, that a dead body still is and make an object of what lies on the dissection table. The surprise is that I can be as well *Vesalius* as well as *Vigevano*, depending on my point of view and on the art of my observership.

The year 1543 – the publication of *De fabrica humani corporis* by Andreas Vesalius – marks the threshold that has been taken by the modern onlooking scientist as far as the human body is concerned. Is it coincidence that in exactly the same year Copernicus published his *De Revolutionibus*? It is the year that the world 'went down' and man was driven from an *geocentric* standpoint to a *heliocentric* standpoint. Driven? Mankind was at the point of mental development to do so. The year 1543 therefore is a *metabletic* year: after that the world will never be what it was before. I strongly disagree with the nowadays broadly accepted idea of Kuhn (Kuhn, 1970) and others, that 1543 has to be considered as the year of a switch of paradigm. With this term we indicate nowadays the shift within the scientific community from the one framework of theories and models (so called paradigm) to another. In my opinion however there happened a lot more that. We all, our culture switched completely from the one reality to the other. In this respect the micro-event on the dissection room may be considered as an recapitulation of what happened to us as mankind, as culture.

Two worlds?

We return to the essential of reductionism. What happened? What happens every day? What is the essential mind related to the *genetic essentialism* of Lindee and Nelkin (Nelkin, 1995), to the attitude of '*DNA-thinking*'? In the decades of Descartes, Bacon and all the others, we lost a horizon. We had to and our success was overwhelming. Like Bacon foresaw, a new reality was revealed: in the extension of objectivism and onlookership a new unknown manipulative substrate was revealed. Wisdom became applicable knowledge. In the extension of the reduction from *given* nature to *thought* nature lies *made or constructed* nature. This does not raise problems for our mentality, unless we confuse the conditions for a given matter for the matter itself! We might confuse two realities and make equal what is not equal at all. For example, I teach my students, ofcourse you can manipulate chemical values, you may destroy bacteria in an diseased body, but look who is healing? Or dying? You always have to deal in the end in essential with the patient, the living body. There is illness, there is pain, there is dying. The body of medical science is a secondary structure, that indeed is a necessary but not sufficient condition for the primary structure, but not the reverse.

That is the sting of the 'bad' reductionism as it is meant here! The world of the *first or primary structure* is the world as it is revealed by our senses and subject. Every modern child still awakes in that world. It is the world where the sun sets down at the horizon. Don't think that the medieval scientist thought that the earth was flat – or at best lightly curved –. He did not think that, he observed it! Like you and I can do if we take the geocentric i.e. primary standpoint, the standpoint of partnership, of *participating observation.* If you literally participate this earth, this is your reality.

A sunset is not an illusion, it is evidence. Nothing wrong with it. Reality? Sure it is! At least if you take your senses and subject seriously! This is what the phenomenologist calls the '*pre-scientific*' world. This is not meant as 'old fashioned' or 'out of date' or even as an historical term. It is meant as an actual reality, the world that is revealed to us before we mentally step out of it, which is an essential prerequisite to be onlooker and analyze it as a object. Science is meant here as modern ('post-cartesian') science, where we *object-think* nature and the world.

Science creates an abstraction (literally) of the primary world and reveals a *secondary* reality. Which is of course also evident. Evidence apparently is not a scientific principle, it is a 'pre-scientific' experience. Everyone can observe order, *Bauplan*, Gestalt in living beings, in organisms as given evidence, not a thought or added reality. Every participating observer can observe it, describe it and analyze it. The organism that is not a passive substrate in a kind of neutral or independent environment but where environment and organism are the one and the same relation. But as soon as you follow the path of reduction, properties and behavior of the organism are thought as permanent traits and, searching for the substrate – you have to, it is the consequence of your approach –, you find cells, chromosomes, molecules and finally you find the order, the plan represented in a genome or DNA-sequence. The DNA-blueprint is the *secondary* outcome of what is a *primary* experience.

Two worlds, both evident, but in a hierarchical relationship to each other. The danger of a 'mental accident' is serious. In the last centuries so much facts and *res extensa* was brought to light that it was of an overwhelming evidence. Brains came up, neurons, proteins, genes and so on. However, in the mind of so many people this secondary world of evidence became the only reality, it obscured the forgotten world of the first structure. Hear it yourself state to your children: 'What you see there, son, is a sunset, but in reality it is the earth globe that is turning and the sun stands still'. Hear me as anatomist 'Okay you feel your pain in your toe, but in reality it is a matter of endorphin metabolism in your brain'. And so on, and so on. I call this 'foaming science'. *Res extensa* reality filled up the world and with a small but essential *metabletic* jump we made the two worlds alike. With the result we lost one, with the result we misthought the world, like you can misstep. Others talk about mistaking 'given' nature for 'thought' nature (H. Verhoog, 1994).

This process of confusion ends up with what Rupert Sheldrake calls *nothing butterism.* Essential is that in the mind of so many people and scientist the primary world has been ruled out as non-reality. Like in the dissection room we metabletically changed reality and now believe that we see the world 'as it really is'. That is the abused 'monism' of Dennett and Dawkins. First we reduce the world to genes and brains and than we state that that is what reality represents. Completely denying the world of the subject, where it is derived from. Do you have genes? I don't. Do you have brains? I don't. I mean not in the first structure of reality. But sure, I have traits, I have thoughts. And sure, a DNA-sequence, a brain somewhere is conditional for those.

What is the nature of the success of this metabletion of human mind and consciousness? The success is 'that it works'. You have heard Bacon. But a belief is not true because it is useful. Every molecular biologist knows what I mean. We say we 'glue' and 'precisely cut' genes, but we all know that in the end we always deal with organisms: to perform our genetic constructing, but always to control whether what we thought we performed, actually has been performed. Always dealing with the organism, even as small as a bacteriophage. But the outcome of our experiments does not prove that our models are representing reality. They might only parallel it and that is quite a different matter. The question of Chargaff becomes urgent: 'Are we imitating living nature or parodying it?' (Chargaff, 1995).

Cartesian schizophrenia

Science as the main source of images legitimating the world 'as it really is', is not non-partial. As philosophical or mental reductionists, so as 'nothing butterists' we have to deal with cartesian (or better: post-cartesian) schizophrenia. Science deals with facts, not with values. It is the duty of every researcher nowadays to become schizophrenic. Schizophrenic enough to live in two worlds or at least with two complementary concepts of the world. The scientific world on the one hand. And the other? No words are good enough to describe it!

This split becomes more and more urgent in the field of ethics. Where to get your ethical references from? From the world 'as it really is'? But what is that? 'We just give nature a hand' is the often heard comment of modern genetic engineers. Which nature? The nature related to DNA? There things in principle are amoral. A gen-ethics is developing nowadays that is without any resistance in direct line of the scientific models of DNA as the 'molecule of life' to name one odd idea. But it is the ethics of a secondary, that is partial world. Where might we find ethical 'hooks' if we do not want to take them from somewhere else (the feared 'sky-hooks') nor want to take them from the nothing buttering context of pseudo-monistic philosophies like that of Dawkins. But we cannot live without ethics. Every day our judgement is asked about new choices, chances and possibilities. And forming a judgement is based upon opinion. Building your opinion is based upon your image of things. And images are generated massively by science and scientists. And doesn't the molecular biologist claim that he describes life, nature and heredity 'as it really is'. Everyone is aware of the fact that directions we choose for mankind and society are based upon how we consider the world 'as it really is'. The power of scientific metaphors in the awareness of the public is enormous.

The corrosive power of DNA

Philosophically we are dealing with a dangerous substance as DNA is concerned. The notion of 'universal acid' is a game loved by philosophers. It is a illusionary substance. An universal acid may dissolve everything. How could it exist if there is no material to construct containers for it, because every material will be dissolved by the acid? DNA has now the status as icon or holy metaphor for living nature as the 'molecule' is the icon for physics. DNA acts like such an universal acid. A lot of clear entities and values vanish at the level of DNA. Some examples might explain that. 'Why talk any longer about species barriers if DNA is the universal substance of life?' 'Why claim difference in biological meaning or ethical value between plants and animals and man, if they all are ruled by the same substance?' 'What is true for the bacteriophage is true for the mouse, for man'. Like the hypothetical universal acid which no container can hold, DNA threatens to solve all the order of the so called classical biology, at least in the minds of DNA-thinkers. With 'classical biology' is not meant here a theoretical model but the biology of the primary and 'given' nature. Where so many values are evident that are obscured by the equalizing power of DNA, at least of the DNA as we think it in the simple models of genetic engineering? Where the polarity of life and dead matter (for instance in respect to dealing with the physical laws of entropy) is a fact. Where the so common notion that the organism relates to the external as some neutral area of condition is a fiction, but where the artificial (for conceptualized) boundary between the 'inside' (organism) and the 'outside' (environment) does not exist. And so on.

The corrosive power of the DNA-thinking is in particular notable in the smoothing away of diversities, differences, qualities, making *metableticall*y zeros equal to ones. Listen to the euphoria when biologists state that the 'same' genes in species one 'orchestrate' the appearance of a (Drosophila) wing and 'control' the shaping of a leg in another species. What is the notion of 'the same'? That is only true on the level of DNA sequences and within the *one way DNA* of the central dogma of molecular biology, no more no less. The so called 'universality' of DNA is true within the small margins of reductionism. As soon as one allows *DNA in context* and has to define a gene as a process (not as a substance) that might reach from the chromosomal level till the food that the animal eats, one also includes the *one way DNA* of the 'DNA-thinkers'. With preservation of the effects and results of your manipulations (if you like), but with an open mind for the context reality of given nature in which your manipulations really appear and meet their meaning and value.

Partnership and dialogue

The 'pre-scientific' world apparently is not only to be subdued to Bacons onlookers approach. Indeed there are two fundamental scientific attitudes, polar indeed but complementary. That is a promising notion, because if I am concerned here about the loss of horizons and worlds, 'making complete' must be a challenging alternative!

- As explained before the *onlooking observation* is the appropriate attitude for the reductionist. It reveals a reality full of causes, that can be experimentally exploited. As side effect there is separation, the world falls apart in fragments that at best have 'interactions'. There you deal with genetic substances and particles. It relates to causal thinking and the *one way DNA.*
- The *participating observer* deals with organism as the presented entities of life as they are given, with individuality and related intrinsic value as the basic entity for psyche and behavior. He also recognizes the biography as the unique entity of human life. It reveals conditions that of course can be manipulated, not by 'forcing' the being but by so to say 'inviting' its intrinsic potencies to change and come along in the desired direction. It relates to conditional thinking and the *DNA in context.*

Indeed on the one side man that abstracts himself from the world as it is experienced, the world of the primary structure, first in mind, next mentally and finally in his action and handling. In the other side the one who wants to finds the roots for his judgement about 'what things, what nature really are' in a participating mentality. So what to do? Not back to the Middle ages, not an alternative flight in methodology opposite to the reductionism. On the contrary, mastering your attitude, mentality and action means to be aware of limitations and to be able – as a person but at least as community – to follow both or more tracks. That will restore our relationship, the bond broken by the post-cartesian cleft may be healed.

What might be the profit? Ethics no longer has to be something that has to be added, originating from another domain as the scientific framework. Nor are we only dependent on the secular ethics that is the continuation of the amoral entities we have split nature into, neither to the utilistic ethics of neo-darwinism war ethics of competing genes alone. In the domain of an organi(ci)stic approach, there is an ethics to gain that is *sachgemäss*, related to the object itself. There exist notions like *intrinsic value* of the organism, of the landscape, of ecological systems, of Gaia herself.

A human world

Are our common senses closed for the enormous abyss that yawns between the genome and the phenotype? Between human genome and human biography? The human genome, as Watson – the father of the DNA stair case and the mental brother of Crick – states, 'the knowledge of which is the ultimate tool for understanding ourselves?' Is that the GNOOTI SEAUTON of future mankind?

'Man becomes himself thanks to and in spite of his genes' and the real level of man-being, the *biography,* is the stage, the realization of it' (van der Bie, 1994).

I have never seen and never will see a 'diseased gene'. Definitions of being ill or healthy and the definition of related values should come from the 'pre-scientific' levels. Disease and health are human conditions of being. They can only be understood from human existence, not from DNA levels for example.

Conclusive remarks

I tried to drag into your awareness the existence of a 'pre-scientific' reality. I tried to defend that taking that reality into account completes our concept of reality. What could this mean for spirituality? By the fact that the concept of reality for many of us has become synonymous with the reality of the reduced world of the secondary structure, we lost contact with values that in the 'pre-scientific' world evidently are present. Convinced materialists like Dawkins, Minsky, Singer always claim that notions like 'creation, spirit, the I of the person, telos' are thought association produced by human mind and lack any root in reality. This is not true for the 'pre-scientific' world. There such notions and entities are present as experiences as evident as the scientific notions of reductionism. Talking with my students about concepts of evolution I always express my wondering that something like 'plan' or 'order' or 'destiny' or 'telos' in evolution has to be denied, since they are so evidently present for every open mind participating observer. I discovered more and more that the edge between matter and spirit, or *mutatis mutandis* between plan and chance is so small, that you only need a small metabletic jump and everything comes in a complete different context.

Spirit, creation, maybe even GOD, have become notions that lay beyond the facts. That is all due to the epistemological separation we made between *res extensa* and *res cogitans.* There exactly lies the win that a participating observership might reveal. In the 'pre-scientific' reality spirit and matter are both present and actual, an *exact* world (literally: made, out of the act) and *inact* world (literally: making, coming to act). I take an example from the (in my opinion fruitless) debate about '*nature* or *nurture*?'. The organism as the appearance of a 'third' dimension shaped by the conditions of nature and nurture, appearing in time on those *per se* necessary but insufficient conditions. An organism does not come out a molecular blueprint, its is the appearance in time and in uniqueness between (thanks and in spite of) the polar tendencies of plasticity and restraint, nature and nurture, outside and inside. We all can have the experience that we

meet in every living creature a *being*. Why not take that seriously? The facts do not prove the inaccuracy of this notion. Only our political unwillingness to dialogue impedes us to meet a partner and so the revelation of a spiritual dimension.
Let us at least deal with DNA but not be ruled by it. We will not find humanity in our genes, but, because of the increasing progress in genetic diagnostics and possibilities of manipulation, we will be increasingly confronted with questions and dilemmas that challenge our humanity. It will bring us to a new consciousness, a real post-cartesian mentality. In my opinion that is what evolution is all about: not biology but consciousness, mentality. That is why I like to deal with the DNA, in order to overcome my and our genes, not to be ruled by them.

References

Amons, R. et al. (1994) Gedanken zur Entwicklung der Vererbungslehre, in Amons, R. et al (eds.) *Genmanipulation an Pflanze, Tier und Mensch; Grundlagen zur Urteilsbildung*, Verlag Freies Geistesleben, Stuttgart.

Bie, G.H. van der (1994), Genmanipulation beim Menschen – ein medizinisch-ethisches Problem, in Amons, R. et al. (eds.) *Genmanipulation an Planze, Tier und Mensch; Grundlagen zur Urteilsbildung*, Verlag Freies Geistesleben, Stuttgart.

Chargaff, E. (1995) *Ein zweites Leben*, Klett-Cotta, Stuttgart, p. 85.

Davenport, R. (1978), *An outline of animal development*, Addison-Wesley Publishing Company, London.

Dawkins, R. (1976) *The selfish gene*, Oxford University Press, Oxford.

Dennett, D.C. (1995) *Darwins Dangerous Idea, Evolution and the Meanings of Life*, Uitgeverij Contact, Amsterdam/Antwerpen.

Holdrege, C. (1996) *Genetics and the manipulation of Life: The forgotten Factor of Context*, Lindisfarne Press, Hudson, New York.

Kuhn, T.S. (1970) *The structure of scientific revolutions*, (second edition), University of Chicago Press, Chicago and London.

Nelkin, D. and Lindee, M.S. (1995) *The DNA Mystique*, Freeman and Company, New York.

Verhoog, H (1994), Reduktionistisches und organisches Denken in der Wissenschaft, in Amons, R. et al. (eds.) *Genmanipulation an Pflanze, Tier und Mensch; Grundlagen zur Urteilsbildung*, Verlag Freies Geistesleben, Stuttgart.

6 DNA in the environment: ecological, and therefore societal, implications

GUENTHER STOTZKY
Microbiologist
Laboratory of Microbial Ecology, New York University, Department of Biology, New York, NY 10003, USA

Abstract
Microorganisms grow in some environments but not in others, even though most environments are exposed to essentially all microorganisms. Some of the physical, chemical, and biological factors that affect establishment and growth are known. In the 1940's, it was discovered in the laboratory under controlled conditions that bacteria transfer genes, not only by conjugal transfer (conjugation), but also, unique among all other organisms, via viruses (transduction) and extracellular DNA (transformation). In the late 1960's, studies began on whether gene transfer contributed to the adaptation of bacteria to changes in their environments, as it had been assumed that mutation was primarily responsible for changes in the genetic composition of bacteria in natural environments. These studies showed that gene transfer occurs in soil – probably the most complex of environments – and that extracellular DNA, either "naked" or in viruses (bacteriophages), became resistant to degradation when bound on surface-active soil particles, such as clay minerals and humic substances, and persisted. Because this extracellular DNA is not expressed in soil, as it is not in a cell and, therefore, is not detected, it was dubbed "cryptic". If the cryptic genes are "novel" (i.e., contain recombinant DNA and are not naturally present in a bacterium), they could pose a hazard to the environment, as the undetectable genes could persist, even after the introduced engineered bacteria disappeared, and then reappear in another host more adapted to soil. Some novel genes resulted in unanticipated adverse environmental effects, emphasizing that the potential effects of genetically engineered organisms must be evaluated in simulations of the environment to which they are to be released. Another potential hazard is the widespread use of transgenic plants containing genes from bacteria that code for proteins toxic to insects. Although these biotoxins are probably less of an environmental hazard than synthetic chemical pesticides, they can accumulate in soil when bound on surface-active particles and may be toxic to non-target beneficial insects and enhance the enrichment of toxin-resistant target insects.

Although there may be many potential benefits from genetic engineering, the release of recombinant DNA to the environment poses potential risks to society, which must be weighed against potential benefits.

Soil as a microbial habitat

Microbial ecologists ask a simple question: why are some microorganisms (e.g. bacteria, fungi, protozoa, algae) present in some environments and not in others? For example, our bodies are exposed to the many diverse species of microbes present in our surroundings, but very few of these species can be isolated from our skin or from our mouth. Moreover, the species in our mouth are different from those on our skin, and the species in our intestinal tract are different from those in our mouth. Why? At the turn of the century, Martinus W. Beijerinck raised the same question, and he concluded that "everything is everywhere, and the milieu selects" (Brock, 1961). For more than 30 years, my laboratory has been interested in what it is in the milieu that selects, and we have studied the various environmental factors that influence the activity, ecology, and

Table 1

Factors affecting the activity, ecology, and population dynamics of microorganisms in natural habitats
Carbon and energy sources
Mineral nutrients
Growth factors
Ionic composition
Available water
Temperature
Pressure
Atmospheric composition
Electromagnetic radiation
pH
Oxidation–reduction potential
Surfaces
Spatial relationships
Genetics of the microorganisms
Interactions between microorganisms

population dynamics of microorganisms (Stotzky, 1974, 1989). Table 1 summarizes some of these factors. The first 13 are primarily chemical and physical factors. Although the genetics of the microorganisms is also important, the nature of the genes is almost a secondary consideration, as in most environments, the non-biological factors are the most influential. For example, if a microorganism is inoculated into a spectrum of soils that differ in physico-chemical characteristics, the microorganism will establish and grow in some soils but not in others, even though the genetics of the microorganism is identical.

The environment in which my laboratory has primarily studied these factors is soil, as it is probably the most complex of microbial habitats. Soil differs from other environments, as it has a high ratio of solids to liquids and, therefore, surfaces are important. Figure 1 depicts our concept of the structure of the soil environment, which is not definitively known. It is composed of numerous individual microhabitats that are comprised of inorganic particles, such as sand and silt, which have essentially no surface activity. These particles are coated, completely or in part, by clay minerals (called clay skins or cutans) that are surface-active particles and are essentially solid-state crystals containing primarily negative charges but also some positive charges (Stotzky, 1986, 1989). Because of their surface activity, the clays retain water against gravity. This

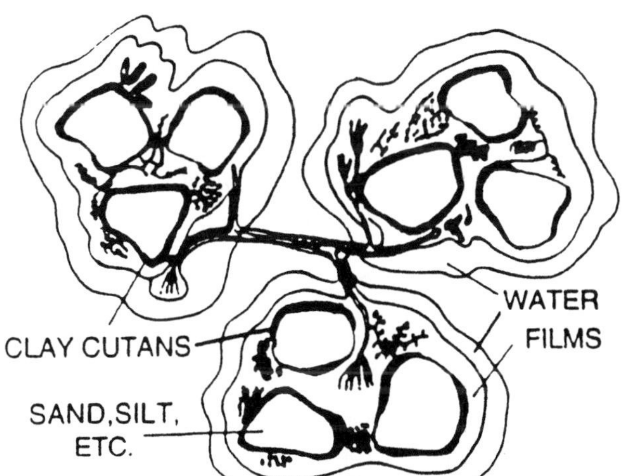

Figure 1. Schematic representation of microhabitats in soil.
The space between each microhabitat is the gas-filled pore space. Note the fungal hyphae growing through the pore space from one microhabitat to another; the hyphae are surrounded by water, which may contain bacterial cells and bacteriophages. (Not to scale.)

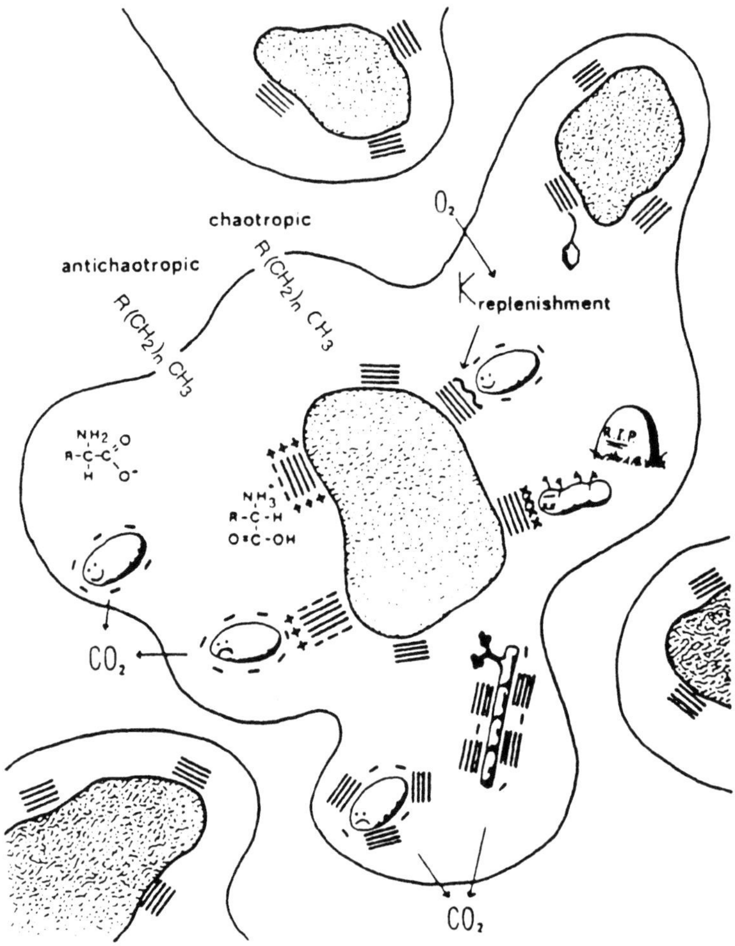

Figure 2. Schematic representation within a microhabitat in soil.

The stippled areas represent sand or silt particles, and the families of adjacent short lines represent packets of clay mineral, with the location of positive and negative charges indicated. The wavy line on one clay packet indicates bound substrates, and the series of x's on the other packet indicates bound substances toxic to microbes. The adsorption of a positively charged protein (i.e., at a pH below its isoelectric point) on negative sites on clay, of a negatively charged bacterium on positive sites on clay, and of a bacteriophage by its negatively charged tail on positive sites on clay is indicated. The physiological state of bacteria in different associations with particles in the microhabitat is indicated. (Not to scale.)

retention is of great importance as, regardless of where they live, microorganisms are aquatic creatures and require water for their metabolism. If clays were not present, soil would drain quickly, similar to sand on a beach, which, even after a heavy rain, is shortly dry. The clay-coated particles with their associated water films form individual microhabitats. The space between the microhabitats (the pore space) is normally occupied by atmosphere containing oxygen, carbon dioxide, and other gases, except when soil is saturated with water.

Because of the discreetness of the microhabitats, it is difficult for microorganisms to move from one to another, as there are normally no connecting water films between the habitats. Thus, microorganisms are generally restricted to individual habitats, although filamentous forms, such fungi, may grow from one habitat to another because they carry their own water films with them. This discreetness restricts the transfer of genes among microorganisms, especially bacteria, in soil.

Even less is known about the interior of a microhabitat, and we can only speculate about its structure and composition. Our concept of a microhabitat in soil is depicted in figure 2. The microorganisms generally live in the water films associated with the solid phase, because if the microorganisms were to attach to a particle, they would depend on diffusion, which is very slow, for replenishment of their nutrients. Free to move around in the water films, they have greater access to nutrients. In contrast, viruses of bacteria – which are called bacteriophages – attach tightly to clay minerals, as does deoxyribonucleic acid (DNA). Consequently, clays are important as surfaces for the binding of inorganic and organic substances (e.g. proteins and DNA, which can be positively or negatively charged depending on conditions within the microhabitats) that can serve as nutrients and genes for microorganisms. Humic substances, which are surface active, are probably also involved in these phenomena (Crecchio and Stotzky, 1997a, b). The importance of these surface interactions will become more obvious later.

Role of DNA in the cell

Until the mid-1940's, the genetic material was assumed to be protein (Table 2). This assumption was reinforced in 1935 when Stanley crystallized the tobacco mosaic virus (TMV) and showed that it was composed of protein (Prescott et al., 1996). A few years after the studies of Stanley, Bawden and Pirie demonstrated that nucleic acid was present in the TMV, but no one paid much attention to this. Another generally accepted concept that predated the mid-1940's was that bacteria did not have sex; i.e., gene transfer and genetic recombination did not occur, and all genetic changes were thought to occur by mutation. This, despite the elegant and simple experiments by Griffith in 1928 that demonstrated that bacteria can transfer some of their traits to other bacteria (Griffith, 1928). When Griffith injected mice with a pathogenic bacterium that causes pneumonia, the mice died, and he was able to recover the bacterium from the dead mice. However, when he injected mutant, non-pathogenic forms of the same bacterium, the mice lived,

Table 2

SOME MILESTONES IN THE HISTORY OF GENE TRANSFER AMONG BACTERIA

• 1935	Stanley crystallized tobacco mosaic virus (TMV); showed protein nature
• 1938	Bawden & Pirie showed presence of nucleic acid (RNA) in TMV
• 1928	Griffith demonstrated transformation ("naked" DNA)
• 1944	Avery, MacLeod, & McCarty demonstrated that DNA was the transforming principle (first demonstration that DNA is the genetic material)
• 1946	Lederberg & Tatum demonstrated conjugation (cell-to-cell contact)
• 1951	Lederberg & Zinder demonstrated transduction (bacteriophage)
• 1972	Weinberg & Stotzky demonstrated conjugation in soil
• 1973	Cohen & Boyer synthesized first recombinant plasmid capable of being replicated in a bacterium
• 1996	Prions and virinos (protein only?)

and he could not recover the bacterium. When he injected the non-pathogenic bacterium together with heat-killed cells of the pathogenic parent strain, the mice died, and he recovered pathogenic bacteria. Griffith called this phenomenon transformation. However, essentially no one paid much attention to these studies. In 1944, Avery and his colleagues at the Rockefeller University demonstrated that the transforming principle was DNA (Avery et al., 1944). This was a milestone! Their studies demonstrated two things: first, that DNA is the hereditary material, which devastated the dogma that it was protein; and second, that genetic recombination can occur in bacteria as in most other organisms. In 1946, Lederberg and Tatum demonstrated another form of gene transfer in bacteria called conjugation (Lederberg and Tatum, 1946), wherein the DNA moves from a donor bacterial cell to a recipient bacterial cell through a conjugation bridge connecting the cells. In 1951, Zinder and Lederberg showed a third form of transfer called transduction (Zinder and Lederberg, 1952), wherein a bacteriophage, which has incorporated a few genes from a bacterial cell that it has infected and killed, transfers the DNA to a newly infected cell without killing that cell.

Although these pioneering studies demonstrated many things, they were done in the laboratory under controlled conditions, and a basic question remained: does gene transfer occur in nature? Several investigators began to study this by looking for conjugation in the intestinal tracts of animals (Stotzky and Babich, 1986). Although they demonstrated transfer, it was not clear whether the transfer occurred in the intestinal tract or outside in the fecal material. Nevertheless, transfer did occur. In the 1970's, my laboratory began to study whether genetic recombination occurs in soil. We showed that conjugation occurs in soil, and, subsequently, we showed that gene transfer by other methods also occurs in soil (Gallori et al., 1994; Khanna and Stotzky, 1992; Krasovsky and Stotzky, 1987; Lee and Stotzky, 1990; Stotzky, 1989; Stotzky and Krasovsky, 1981; Stotzky et al., 1990, 1991, 1993, 1996; Vettori et al., 1996; Weinberg and Stotzky, 1972; Zeph et al., 1988).

Another important milestone occurred in 1973, when Cohen and Boyer synthesized the first recombinant plasmid (Cohen, 1975). A plasmid is a small piece of circular, extrachromosomal DNA, i.e., it is not part of the bacterial chromosome. Because enzymes that could cleave DNA at specific sites already had been discovered, Cohen and Boyer were able to alter a plasmid by inserting a piece of DNA from another source. This synthesis of the first recombinant plasmid really began what is now called the biotechnology revolution.

To emphasize that dogmas do not die easily and that new dogmas may be incorrect or incomplete, infectious particles exist that are smaller than viruses and seem to consist only of protein (Prescott et al., 1996). These particles are called prions and virinos, and as they are apparently also capable of transferring genetic information, nucleic acids are presumably present, although this has not been conclusively demonstrated. Prions and virinos are important because of the serious diseases that they cause, e.g., "mad-cow" disease, Creutzfeldt-Jacob disease, possibly Alzheimer's disease.

The ability to construct recombinant DNA and insert it into microorganisms began to worry some scientists, environmentalists, and people in regulatory agencies. They wondered what would happen if recombinant organisms are released to the environment (Doyle et al., 1995). My laboratory was interested in this question, as it was viewed as an extension of our earlier studies on gene transfer in soil, although those earlier studies were not motivated by genetic engineering but by whether or not gene transfer in the environment was important for the adaptation and evolution of bacteria. We were successful in obtaining funds from the US Environmental Protection Agency to study the survival and effects of genetically engineered microorganisms (GEMs) introduced to soil.

GEMs in the environment

Some of the questions that we and others raised about the potential release of GEMs to the environment are summarized in Table 3. "Survival" does not mean that the GEMs are just "sitting" there, for if they just sit and do nothing, they will not efficaciously perform the functions for which they have been engineered and released. Rather, they have to

Table 3

Questions That Must Be Answered in Assessing the Risks Involved in the Release of Genetically Engineered Microbes (GEMs) to the Environment
Survival of the GEMs
-Establishment (colonization)
-Multiplication
Dispersal of the GEMs to other sites and environments
Transfer of the novel genetic material to indigenous microbes
-Survival
-Establishment
-Multiplication
Potential ecological and health impacts of the novel genetic material
Probability of containment, decontamination, and mitigation

establish and multiply. Another question is: if GEMs are introduced to soil, will they eventually end up elsewhere, such as in surface and ground water? Another question is: does gene transfer occur from the introduced GEMs to the naturally-occurring, indigenous microorganisms in soil, i.e., so called horizontal transfer? If so, do these new GEMs survive, establish, and multiply? The probability of this is greater than with the introduced GEMs, as the indigenous microbes are obviously adapted to the soil environment.

These are interesting questions, but, in a sense, they are academic questions. The "bottomline" question is: are there any potential adverse ecological or health effects resulting from the release of GEMs? If the GEMs have no detrimental effects and just do the job for which there were designed, everything would be fine. However, concern

Table 4

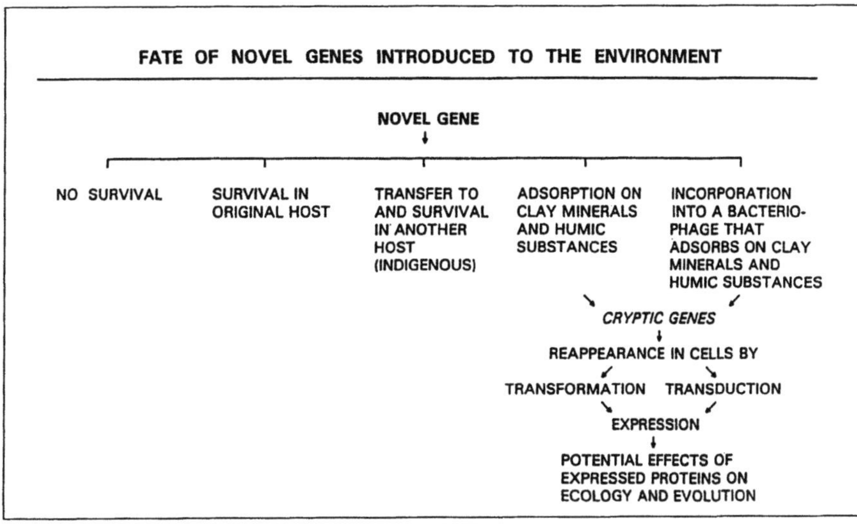

arises if there are potential adverse effects. This leads to the related concern of whether it would be possible to decontaminate an environment if the released GEMs had an adverse effect. This could be a major problem, as this would not be the same as recalling 100,000 automobiles because of faulty brakes. Once the GEMs are released to the environment and they establish, they are out there to stay! This is an important consideration, because the experience with the release of some organisms into new environments has not always been positive (e.g., kudzu grass in the United States; rabbits in Australia).

What might happen to the novel genes in GEMs released to the environment? As shown in Table 4, the genes may not persist and simply disappear. They may remain in the original host bacteria and survive if the host survives. They may also be transferred to other microorganisms and persist there, especially if these new hosts are indigenous bacteria. On the other hand, the original or new hosts may die and release the genes to the environment, where the extracellular DNA (called "naked" DNA) may adsorb on surface active particles, such as clay minerals and humic substances, and be protected against degradation. The DNA may also be incorporated into bacteriophages, which, after being released from killed cells, may adsorb on clay minerals and humic substances and be protected against inactivation. In the latter two cases, the genes are there, even though they are not expressed, as they are not present in cells. These genes are called "cryptic" genes. However, the cryptic genes may not always remain dormant. The naked DNA and the DNA in bacteriophages may reappear in bacteria via transformation or transduction, respectively. If the information encoded on a gene in a new host is

Table 5

Survival of GEMs in Soil

- reduced by presence of indigenous microbiota
- minimum of 30 days (some <30 days)
- linear decrease of inoculum by 2 to 3 orders of magnitude
- independent of plasmid size (2.6 to 64 Mdal)
- dependent on host bacterium
- independent of nutrient additions (initially or after inoculation)
- enhanced by montmorillonite but not by kaolinite
- partitioning (loss of plasmid) dependent on growth
- presence of plasmid sometimes enhanced
- debilitation: "viable but nonculturable"
- chimeric DNA (from Drosophila grimshawi) had little effect

transcribed (all that DNA does is to provide information for the synthesis of proteins), the proteins eventually formed could affect the ecology of other microbes and the evolution of bacteria in that environment.

The survival of GEMs in soil is a function of many factors. For example, as shown in Table 5, survival, in this case of genetically engineered strains of Escherichia coli, the bacterium most used in genetic studies but not a normal inhabitant of natural environments other than of the intestinal tract of many animals, was reduced by the presence of indigenous microorganisms; i.e., in sterilized soil, the introduced GEMs can persist indefinitely. However, in non-sterile soil, competition with and amensalism by other organisms rapidly reduces the number of added GEMs; e.g., within 30 days or so, there was a decrease of 2 to 3 orders of magnitude, and eventually the GEMs and their novel genes could not be detected. The species or subspecies of the host bacteria but not the size of the introduced plasmid appeared to be important. The addition of nutrients,

Table 6

METHODS OF TRANSFER OF GENETIC MATERIAL IN BACTERIA
Conjugation (cell to cell contact)
Transduction (via a bacteriophage)
Transformation (uptake of "naked" DNA)

the type of clay minerals in the soil, and other factors have variable effects on the survival of GEMs (Devanas and Stotzky, 1986, 1988; Devanas et al., 1986).

As mentioned earlier, transfer of DNA, both chromosomal and plasmid, among bacteria can occur primarily in three different ways (Table 6). Conjugation involves cell-to-cell contact. In transduction, the DNA from one cell is incorporated into a bacteriophage, which can transfer the DNA to other susceptible bacteria. In transformation, extracellular

Table 7

FREQUENCY OF GENE TRANSFER BY GENETICALLY MODIFIED MICROORGANISMS IN SOIL

- CONJUGATION - CHROMOSOMAL GENES: 10^{-5} TO 10^{-4}
  - PLASMID-BORNE GENES: 10^{-7} TO 10^{-4}
- TRANSDUCTION: 10^{-5} TO 10^{-1}
- TRANSFORMATION: 10^{-7} TO 10^{-5}

APPEARANCE OF FIRST ORGANISMS ON EARTH: $\sim 3.5 \times 10^{9}$ YEARS AGO;

$\therefore$, ASSUMING AN AVERAGE TRANSFER FREQUENCY OF $\sim 10^{-5} \times 3.5 \times 10^{9}$ YEARS $\simeq 3.5 \times 10^{4}$ TRANSFERS

(VERY CONSERVATIVE ESTIMATE!!)

DNA can be taken up directly from the environment by susceptible (so called competent) bacteria.

These three forms of gene transfer have been demonstrated in almost all habitats, including soil, although the frequency of transfer is not very high (Yin and Stotzky, 1997). For example, as depicted in Table 7, transformation in soil may involve only 1 bacterium in 100,000 to 1 in 10,000,000 (Stotzky, 1989). These are very small numbers. However, when the total numbers of organisms in soil are considered – there may be a billion bacteria per gram of soil – these numbers are not so small. Moreover, if the time for the appearance of the first microorganisms on Earth at about 3.5 billion years ago is accepted, there would have been somewhere around 35,000 transfers during this period. This is obviously a conservative estimate, as it assumes only one successful transfer per year over the entire Earth. Consequently, the probability of gene transfer appears to be very high. However, it must be emphasized that the frequencies of transfer shown in Table 7 were not observed in soil in situ but were measured in experiments in which both donor DNA and appropriate recipient bacteria were introduced to soil. The transfer of genes to and among indigenous microorganisms in soil appears to be very low, if it occurs at all, and only a few investigators have reported such transfer. Hence, even though experimental data suggest that there may be a high probability of gene transfer, in reality it is probably low. Only if gene transfer is considered in terms of geological time, do the frequencies appear to be large enough to indicate that such transfers have significantly affected the evolution of microorganisms.

Effects of GEMs in soil

What are the potential risks of GEMs in the environment? To study this question, we used a bacterium called Pseudomonas putida PPO301 that carried a recombinant plasmid designated pRO103 (Doyle et al., 1991; Short et al., 1991). This plasmid contained genes for the degradation of the herbicide, 2,4-dichlorophenoxyacetic acid (2,4-D), which does not normally degrade quickly in soil. When P. putida contains this plasmid, 2,4-D is

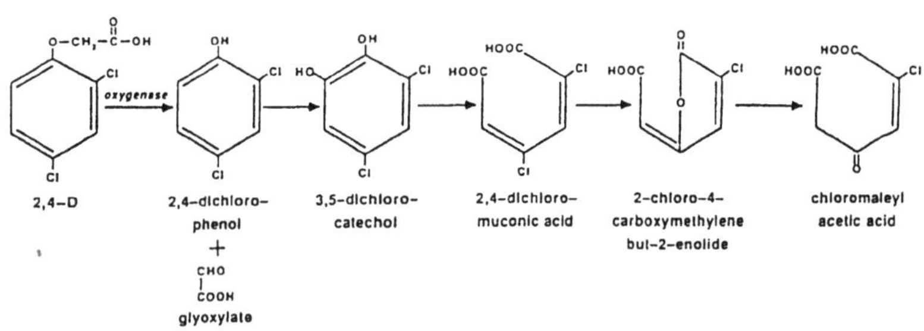

Figure 3.

Plasmid pRO103, mediates the degradation of 2,4-D to chloromaleylacetate by the pathway shown.

converted to chloromaleyl acetic acid (CAA), a harmless chemical that is excreted from the cells and is rapidly converted to carbon dioxide, water, and chloride by many microorganisms in soil (Figure 3). The engineered P. putida [designated as P. putida PPO301(pROI03)] derives no energy from the reaction, because it does not mineralize the 2,4-D to carbon dioxide and water but only converts it to CAA, and it disappears after the 2,4-D has been depleted. This is a simple and effective way to utilize and control this GEM.

The 2,4-D-degrading bacterium was added to soil, and the total microbial activity in the soil was measured by determining the respiration of the soil (by the amount of carbon dioxide produced) over time (Stotzky et al., 1993). As the soil used had a low content of carbon and energy sources, glucose (a simple sugar) was added to some of the soil to enhance the growth of microorganisms. In Figure 4, the upper set of curves shows the respiration of the control soil to which 2,4-D but no bacteria were added. The middle set of curves summarizes what happened when the parental organism (P. putida PPO301), which did not contain plasmid pROI03 that enabled degradation of 2,4-D, was introduced. The results were similar to those obtained with the uninoculated control soil; i.e., the presence of 2,4-D had essentially no effect on respiration, either in the presence or absence of glucose. However, when the GEM containing the novel genes [i.e., P. putida PPO301(pROI03)] was added to the soil, there was a dramatic reduction for about 30 days in the amount of carbon dioxide evolved from the soil containing both glucose and 2,4-D (Figure 4, bottom set of curves), indicating a reduction in microbial activity in the soil. In terms of geological time, 30 days is not a significant amount of time, but these were laboratory experiments in which time is compressed, and the reduction shows the possibility of a significant effect on the soil microbiota if this GEM was released to the environment.

How did this GEM reduce the microbial activity? A partial answer was provided by analyses of the numbers of fungal propagules in the soil. The top and middle sets of curves in Figure 5 show that the numbers of the propagules remained essentially constant in both the uninoculated soil and in soil inoculated with the parental bacterial strain, regardless of the addition of 2,4-D or glucose, indicating that the parental strain (i.e., without the novel plasmid genes) did not affect fungal population. However, as shown in the lower set of curves in Figure 5, when the GEM was introduced into soil containing 2,4-D, there was a rapid reduction in the numbers of fungal propagules. After about 20 days, fungi could no longer be detected, and they did not reappear during the 50-day duration of the study. A similar reduction in the numbers of bacteria did not occur, indicating that the combination of 2,4-D and the GEM affected primarily the fungal component of the soil microbiota.

When the soils were analyzed for the disappearance of 2,4-D and the appearance of the intermediates in the degradation pathway of 2,4-D (see Figure 3) by gas chromatography-mass spectrometry, the 2,4-D persisted in both the uninoculated soil and in the

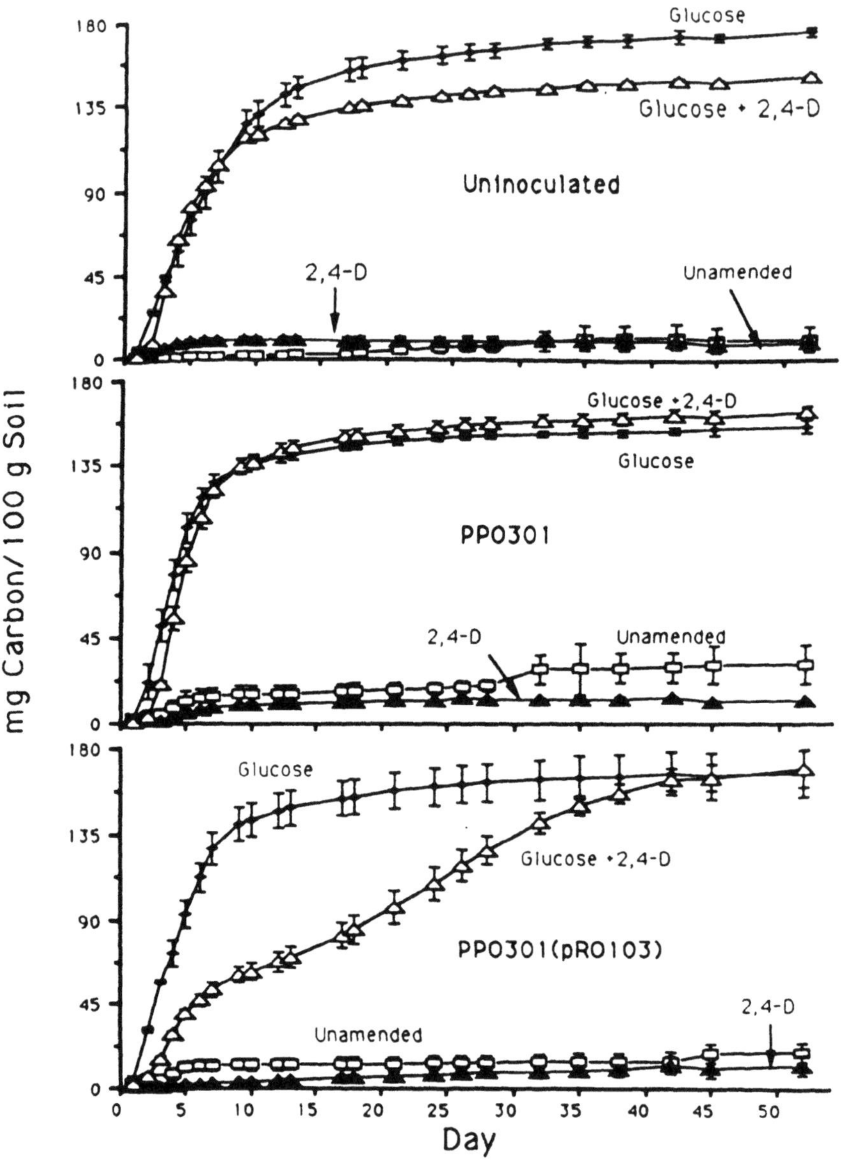

Figure 4.

Cumulative amount of carbon evolved as CO_2 during a 53-day incubation from soil inoculated with *P. putida* strains PPO301(pRO103) or PPO301 or uninoculated and amended with glucose, 2,4-D, glucose + 2,4-D, or unamended. Plotted values are means ± the standard errors of the means.

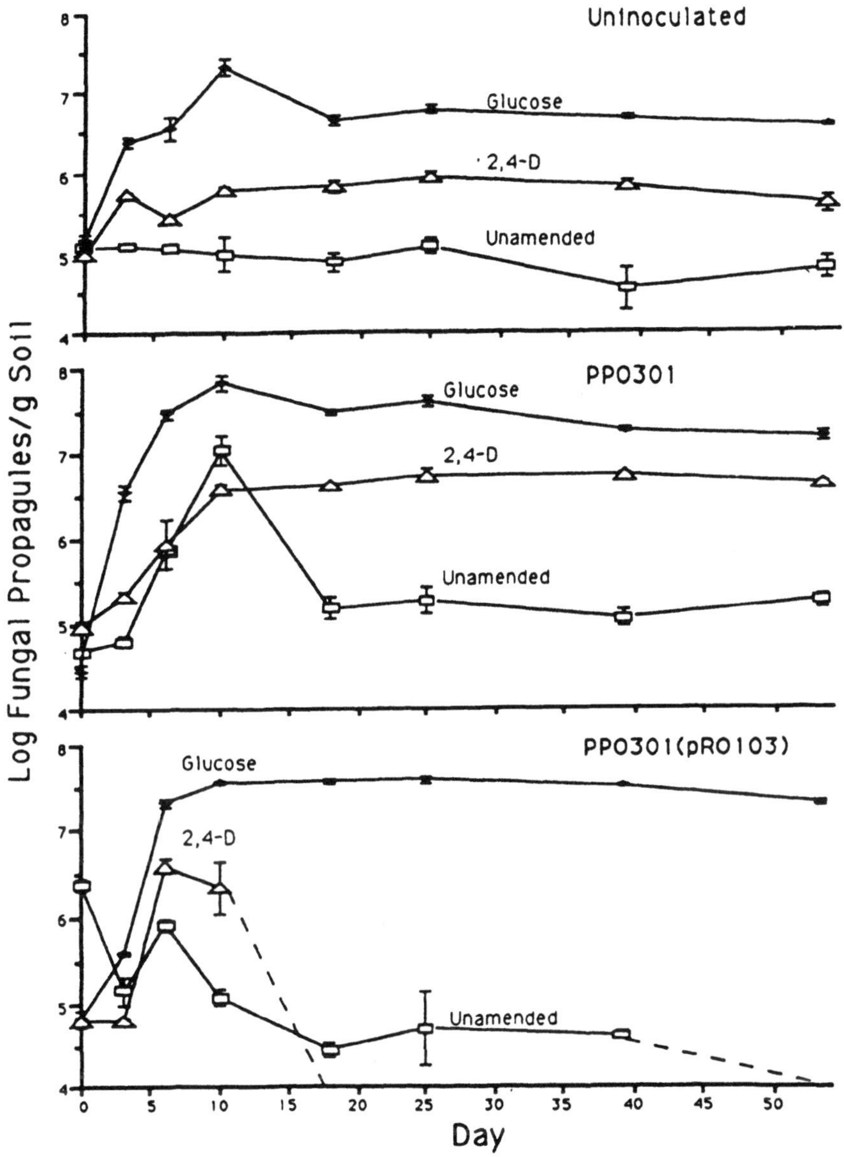

Figure 5.
Numbers of fungal propagules detected during a 53-day incubation in soil inoculated with *P. putida* strains PPO301(pRO103) or PPO301 or uninoculated and amended with glucose or 2,4-D or unamended. Plotted values are means ± the standard errors of the means.

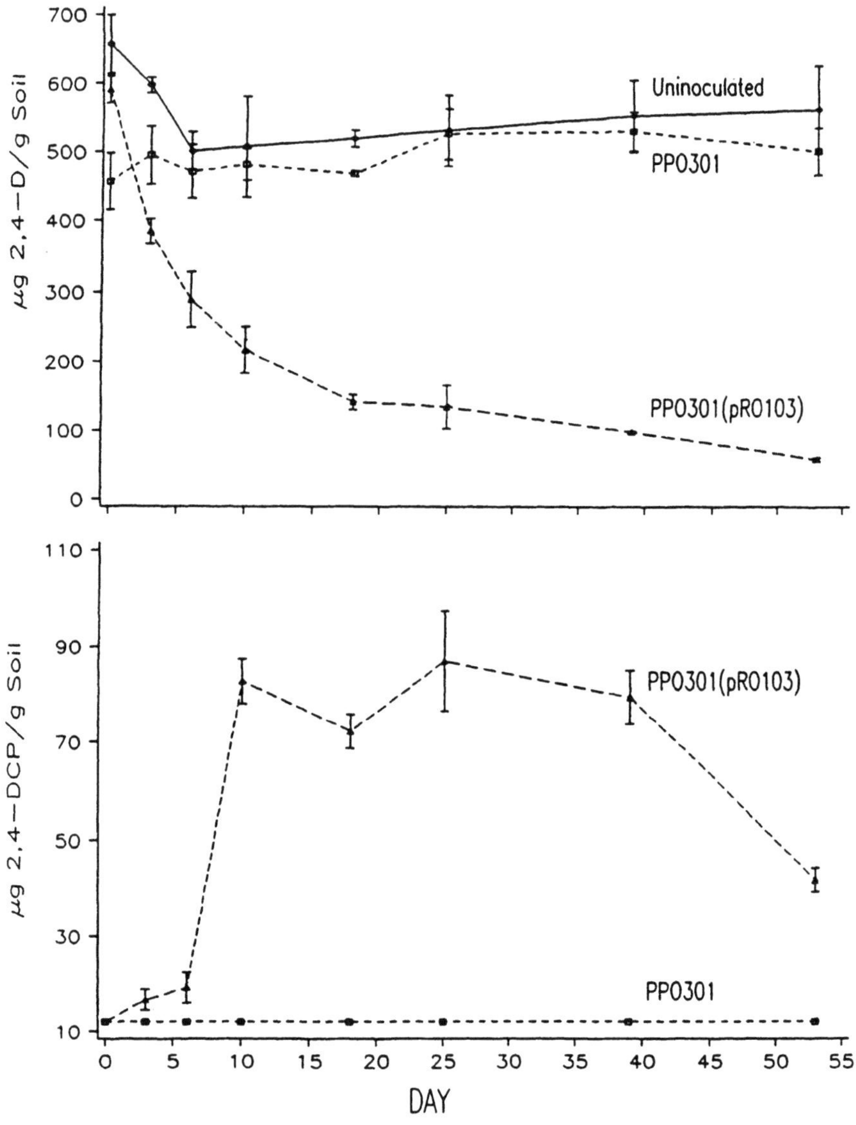

Figure 6.

(A) Kinetics of 2,4-D degradation in soil amended with 2,4-D and inoculated with *P. putida* strains PPO301(pRO103) or PPO301 or uninoculated. (B) Accumulation of 2,4-DCP in 2,4-D-amended soil inoculated with PPO301(pRO103) or PPO301. Plotted values have been corrected for extraction efficiency and represent the means ± the standard errors of the means.

soil inoculated with the parental organism. However, when the GEM was present, the 2,4-D disappeared (Figure 6). The GEM did what was expected: it degraded the 2,4-D. However, it initially only converted the 2,4-D to 2,4-dichlorophenol (2,4-DCP), which accumulated in the soil. In the uninoculated soil or in the soil inoculated with the parental strain, there was no production of 2,4-DCP, as there was no degradation of 2,4-D. 2,4-DCP is very toxic to organisms: e.g., only about 50 parts per million (ppm) of 2,4-DCP inhibits the growth of fungi. Consequently, the inhibition of microbial activity and fungi observed in this experiment was the effect of a delayed conversion of 2,4-D to harmless CAA. Instead, the 2,4-D was rapidly converted into the first intermediate in the degradation pathway, 2,4-DCP, which accumulated, as the enzyme required for its conversion, 2,4-DCP hydroxylase, is under the control of the product of a different gene. Therefore, 2,4-DCP accumulated until this enzyme was induced and further conversion to CAA eventually occurred.

These data indicate that if a GEM is to be released to the environment, it must be thoroughly tested first in simulations of the environment to which it will be released. This testing must be done on a case-by-case basis. The current tendency is to lump organisms into generic groups, and if one member of the group shows no problem, there is presumably no need to test other members. This is an imperfect approach, as these studies illustrate. Even a theoretically "perfect" GEM (e.g., one that converts an environmental toxicant to harmless end-products and then disappears when the toxicant is gone) may affect the environment in unexpected and negative ways. Hence, it should be obvious that GEMs should be evaluated on a case-by-case basis.

Transgenic plants

Another aspect of the potential ecological effects of the products of novel DNA in the environment relates to transgenic plants. Examples are genes that code for the insecticidal toxins produced by subspecies of the bacterium Bacillus thuringiensis (Bt) and that have been engineered into plants (Höfte and Whiteley, 1989). There are numerous types of Bt toxins, and they are highly insecticidal to the larvae of different but specific groups of insects. The normal genes in Bt code for protoxins, which are large proteins that are inactive as insecticides. They are present in crystalline forms and are converted to active toxins after ingestion by insect larvae (Table 8). The guts of insect larvae are usually alkaline (a pH of about 10.5 to 11). This high pH is necessary for the solubilization of the protoxins, which are then cleaved into active toxins by proteolytic enzymes in the gut. The active toxin molecules bind to specific receptors on the epithelium lining of the larval gut, which changes the permeability of the epithelial cells, causes a shift in cellular sodium and potassium levels, and the insect stops eating and starves to death.

However, genetic engineers have introduced incomplete (truncated) versions of the Bt toxin genes into plants. These truncated genes do not code for the complete crystalline

Table 8

ENVIRONMENTAL ASPECTS OF NORMAL AND GENETICALLY-ENGINEERED INSECTICIDAL TOXINS FROM *BACILLUS THURINGIENSIS*

NORMAL GENES	ENGINEERED GENES (TRUNCATED)
code for ↓	code for ↓
PROTOXINS (INACTIVE)	TOXINS (ACTIVE)

↘ ingestion by larvae ↙

↓

LARVAL GUT

↙	↘
• solubilization (pH > 10.5) • cleavage to active toxins by specific proteolytic enzymes • binding to specific receptors on gut epithelium	• already solubilized • already active toxins (no cleavage required) • receptors present also in nontarget insects

↘ ↙

- change in permeability of epithelial cells
- efflux of potassium
- cessation of feeding
- DEATH

FATE AND EFFECTS IN THE ENVIRONMENT

↓	↓
• limited binding and, therefore, rapid degradation and little accumulation • any accumulated protoxins must be processed in gut, as above • specificity for target organisms • high and repeated applications of commercial preparations can result in selection of toxin-resistant target organisms	• rapid and strong binding on clay minerals and humic substances → accumulation • bound toxins resistant to biodegradation and retain insecticidal activity • enhancement of selection of toxin-resistant target organisms • hazard to nontarget organisms

(Are soil organisms resistant to the toxins because of the constant presence of the toxins in soil?)

protoxins (which would be difficult to be translocated in the plants), but rather for the active soluble form of the corresponding toxins. Consequently, when insect larvae ingest the plant, there is no need for the toxin to be activated in the larval gut; i.e., an alkaline pH and proteolytic enzymes are not necessary, as the protein is already in its active form. Therefore, two of the requirements for toxicity are negated (i.e., a high gut pH and appropriate proteolytic enzymes), and non-target organisms could be susceptible to the active toxins, as receptors for the toxins are apparently also present in non-target insects, albeit in lower numbers than in target insects.

Another consideration is the fate of the Bt toxins in soil, as only the commercially useful portions of transgenic plants will be harvested, and the remainder of the plant biomass containing the toxins will be incorporated into soil. One possibility is that the toxins do not accumulate in the environment, as proteins are readily degraded by indigenous microorganisms. However, the toxins may bind on clay minerals and humic substances and become resistant to biodegradation while retaining their insecticidal activity (Crecchio and Stotzky, 1997; Koskella and Stotzky, 1997; Tapp and Stotzky, 1995, 1997; Tapp et al., 1994; Venkateswerlu and Stotzky, 1992). Consequently, high concentrations of active toxins may accumulate in the environment, which could result in an increase in the selection and enrichment of target insects that have become resistant to the toxins. Moreover, because the accumulated toxins are the active toxins, they could be hazardous to non-target organisms, such as beneficial insects and other organisms at different trophic levels. These aspects have been insufficiently studied. However, evidence is accumulating that repeated high applications of even protoxins are resulting in the selection of toxin-resistant target species in the field (Tabashnick, 1994). This selection could be exacerbated by the widespread planting of transgenic plants, especially if the content of toxins in the plant tissue is too low and the toxins accumulate in soil (Kaiser, 1996). The potential increase in the numbers of toxin-resistant insects could compromise the entire technology of spraying Bt bacteria and their toxins to control insects. This would be especially detrimental to organic farmers who shun the use of chemical insecticides.

Conclusions

These studies with DNA and Bt toxins address the question of how biomolecules interact with surface active particles and how these interactions affect the persistence and biological activity of the biomolecules in the environment (Table 9). Biomolecules, including proteins, DNA, and viruses, are protected against biodegradation and inactivation when bound on soil particles. Thus, they can persist and function in the environment. Binding is important, as the free biomolecules are very rapidly degraded and inactivated. Bound DNA retains its ability to transform bacteria; bound bacteriophages are able to transduce DNA, and they and bound mammalian viruses can kill their host cells; bound toxins from Bt kill insect larvae; and bound enzymes, which are proteins, retain their catalytic activity, although the activity may be reduced (Calamai et al., 1997).

When studying a piece of DNA, as well as the protein products coded by the DNA, in the environment, there needs to be concern about the perception (detection), assessment, and management of the biomolecules. Not only does the scientific community have the

Table 9

SUMMARY OF EFFECTS OF INTERACTIONS OF BIOMOLECULES WITH SURFACE-ACTIVE PARTICLES IN THE ENVIRONMENT
• BIOMOLECULES (E.G., DNA, VIRUSES, PROTEINS) BIND RAPIDLY AND TIGHTLY ON CLAY MINERALS AND HUMIC SUBSTANCES
• BOUND BIOMOLECULES ARE PROTECTED AGAINST DEGRADATION AND INACTIVATION
• BOUND BIOMOLECULES RETAIN BIOLOGICAL ACTIVITY - BOUND DNA TRANSFORMS - BOUND BACTERIOPHAGES TRANSDUCE - BOUND VIRUSES LYSE HOST CELLS - BOUND INSECTICIDAL TOXINS FROM <u>BACILLUS THURINGIENSIS</u> KILL INSECT LARVAE - BOUND ENZYMES CONVERT SUBSTRATES

responsibility for this, but everyone does. Although these processes are primarily the responsibility of scientists, the non-scientific community has an important role in monitoring the processes to make certain that they are objective. Academic scientists are usually objective, but scientists working for industry may not always be as objective, as they also have responsibilities to their employer. All science should ultimately be monitored by society. Consequently, when dealing with DNA, especially recombinant DNA, in the environment, it is the responsibility of the entire community to continuously ask: do we want to release this recombinant DNA to the environment; what are the risks and benefits; are the benefits worth the risks? Perhaps society should say more often: let us pause and consider these releases a bit more and obtain more data before releasing.

References

Avery, O.R., MacLeod, C.M., and McCarty, M. (1944) *Studies on the chemical nature of the substance inducing transformation in pneumococcal types*, J. Exp. Med. 79, pp. 137-159.

Brock, T.D. (1961) *Milestones in Microbiology*, Prentice-Hall, Englewood Cliffs, NJ.

Calamai, L., Ristori, G.G., Fusi, P., and Stotzky, G. (1997) *Interaction of catalase with montmorillonite saturated with inorganic or organic cations: effect on enzymatic activity*, Soil Biol. Biochem. (in press).

Cohen, S.N. (1975) *The manipulation of genes*, Sci. Am. **233**, pp. 25-33.

Crecchio, C. and Stotzky, G. (1997a) *Insecticidal activity and biodegradation of the toxins from Bacillus thuringiensis subsp. kurstaki bound on soil humic acids*, Soil Biol. Biochem. (submitted).

Crecchio, C. and Stotzky, G. (1997b) *Transformation of Bacillus subtilis by DNA bound on humic acids and effect of DNase on the transforming ability of bound DNA*, Soil Biol. Biochem. (submitted).

Devanas, M.A. and Stotzky, G. (1986) *Fate in soil of a recombinant plasmid carrying a Drosophila gene*, Curr. Microbiol. **13**, pp. 279-283.

Devanas, M.A. and Stotzky, G. (1988) *Survival of genetically engineered microbes in the environment*, J. Soc. Ind. Microbiol. **29**, pp. 287-296.

Devanas, M.A., Rafaeli-Eshkol, D., and Stotzky, G. (1986) *Survival of plasmid-containing strains of Escherichia coli in soil: effect of plasmid size and nutrients on survival of hosts and maintenance of plasmids*, Curr. Microbiol. **13**, pp. 269-277.

Doyle, J.D., Short, K.A., Stotzky, G., King, R.J., Seidler, R.J., and Olsen, R.H. (1991) *Ecologically significant effects of Pseudomonas putida PPO301 (PRO103), genetically engineered to degrade 2,4-dichlorophenoxyacetate, on microbial populations and processes in soil*, Can. J. Microbiol. 37, pp. 682-691.

Doyle, J.D., Stotzky G., McClung, G., and Hendricks, C.W. (1995) *Effects of genetically engineered microorganisms on microbial populations and processes in natural habitats*, Adv. Appl. Microbiol. **40**, pp. 237-287.

Gallori, E., Bazzicalupo, M., Dal Canto, L., Fani, R., Nannipieri, P., Vettori, C., and Stotzky, G. (1994) *Transformation of Bacillus subtilis by DNA bound on clay in non-sterile soil*, FEMS Microbiol. Ecol. **15**, pp. 119-126.

Griffith, F. (1928) *Significance of pneumococcal types*, J. Hyg. 27, pp. 113-159.

Höfte, H. and Whiteley, H.R. (1989) *Insecticidal crystal proteins of Bacillus thuringiensis*, Microbiol. Rev. **53**, pp. 242-255.

Kaiser, J. (1996) *Pests overwhelm Bt cotton crop*, Science 273, p. 423.

Khanna, M. and Stotzky, G. (1992) *Transformation of Bacillus subtilis by DNA bound on montmorillonite and effect of DNase on the transforming ability of bound DNA*, Appl. Environ. Microbiol. **58**, pp. 1930-1939.

Koskella, J. and Stotzky, G. (1997) *Microbial utilization of free and clay-bound insecticidal toxins from Bacillus thuringiensis and insecticidal activity after incubation with microbes*, Appl. Environ. Microbiol. (submitted).

Krasovsky, V.N. and Stotzky, G. (1987) *Conjugation and genetic recombination in Escherichia coli in sterile and non-sterile soil*, Soil Biol. Biochem. **19**, pp. 631-638.

Lederberg, J. and Tatum, E.L. (1946) *Gene recombination in E. coli*, Nature **158**, p. 558.

Lee, G.W. and Stotzky, G. (1990) *Transformation is a mechanism of gene transfer in soil*, Korean J. Microbiol. **28**, pp. 210-218.

Prescott, L.M., Harley, J.P., and Klein, D.A. (1996) *Microbiology*, Wm. C. Brown, Dubuque, IA.

Short, K.A., Doyle, J.D., King, R.J., Seidler, R.J., Stotzky, G., and Olsen, R.H. (1991) *Effects of*

2,4-dichlorophenol, a metabolite of a genetically engineered bacterium, and 2,4-dichlorophenoxy-acetate on some microorganism-mediated ecological processes in soil, Appl. Environ. Microbiol. **57**, pp. 412-418.

Stotzky, G. (1974) Activity, ecology, and population dynamics of microorganisms in soil, in A.I. Laskin and H. Lechevalier (eds.), *Microbial Ecology,* Chemical Rubber Co., Cleveland, OH, pp. 57-135.

Stotzky, G. and Krasovsky, V.N. (1981) Ecological factors that affect the survival, establishment, growth, and genetic recombination of microbes in natural habitats, in S.B. Levy, R.C. Clowes, and E.L. Koenig (eds.), *Molecular Biology, Pathogenicity, and Ecology of Bacterial Plasmids,* Plenum Press, NY, pp. 31-42.

Stotzky, G. (1986) Influence of soil mineral colloids on metabolic processes, growth, adhesion, and ecology of microbes and viruses, in P.M. Huang and M. Schnitzer (eds.), *Interactions of Soil Minerals with Natural Organics and Microbes, Soil Science Society of America,* Madison, WI, pp. 305-428.

Stotzky, G. and Babich, H. (1986) *Survival of, and genetic transfer by, genetically engineered bacteria in natural environments,* Adv. Appl. Microbiol. **31**, pp. 93-138.

Stotzky, G. (1989) Gene transfer among bacteria in soil, in S.B. Levy and R.V. Miller (eds.), *Gene Transfer in the Environment,* McGraw-Hill, NY, pp. 165-222.

Stotzky, G., Devanas, M.A., and Zeph, L.R. (1990) *Methods for studying bacterial gene transfer in soil by conjugation and transduction,* Adv. Appl. Microbiol. **35**, pp. 57-169.

Stotzky, G., Gallori, E., and Khanna, M. (1996) Transformation in soil, in A.D.L. Akkermans, J.D. van Elsas, and F.J. de Bruijn (eds.), *Molecular Microbial Ecology Manual,* Kluwer Academic Publishers, Dordrecht, The Netherlands, 5.1.2, pp. 1-28.

Stotzky, G., Zeph, L.R., and Devanas, M.A. (1991) Factors affecting the transfer of genetic information among microorganisms in soil, in L.R. Ginzburg (ed.), *Assessing Ecological Risks of Biotechnology,* Butterworth-Heinemann, Stoneham, MA, pp. 95-122.

Stotzky, G., Broder, M.W., Doyle, J.D., and Jones, R.A. (1993) *Selected methods for the detection and assessment of ecological effects resulting from the release of genetically engineered microorganisms to the terrestrial environment,* Adv. Appl. Microbiol. **38**, pp. 1-98.

Tabashnick, B.E. (1994) *Evolution of resistance to Bacillus thuringiensis,* Ann. Rev. Entomol. **39**, pp. 47-79.

Tapp, H., Calamai, L., and Stotzky, G. (1994) *Adsorption and binding of the insecticidal proteins from Bacillus thuringiensis subsp. kurstaki and subsp. tenebrionis on clay minerals,* Soil Biol. Biochem. **26**, pp. 663-679.

Tapp, H. and Stotzky G. (1995) *Insecticidal activity of the toxins from Bacillus thuringiensis subsp. kurstaki and subsp. tenebrionis adsorbed and bound on pure and soil clays,* Appl. Environ. Microbiol. **61**, pp. 1786-1790.

Tapp, H. and Stotzky, G. (1997) *Persistence of the insecticidal toxins from Bacillus thuringiensis subsp. kurstaki in soil,* Soil Biol. Biochem. (submitted).

Venkateswerlu, G. and Stotzky, G. (1992) *Binding of the protoxin and toxin proteins of Bacillus thuringiensis subsp. kurstaki on clay minerals,* Curr. Microbiol. **25**, pp. 225-233.

Vettori, C., Paffetti, D., Pietramellara, G., Stotzky, G., and Gallori, E. (1996) *Amplification of bacterial DNA bound on clay minerals by the random amplified polymorphic DNA (RAPD) technique*, FEMS Microbiol. Ecol. 20, pp. 251-260.
Weinberg, S.R. and Stotzky, G. (1972) *Conjugation and genetic recombination of Escherichia coli in soil*, Soil Biol. Biochem. 4, pp. 171-180.
Yin, X. and Stotzky, G. (1997) *Transfer of genetic information among bacteria in soil*, Adv. Appl. Microbiol. (in press).
Zeph, L.R., Onaga, M.A., and Stotzky, G. (1988) *Transduction of Escherichia coli by bacteriophage P1 in soil*, Appl. Environ. Microbiol. 54, pp. 1731-1737.
Zinder, N.D. and Lederberg, J. (1952) *Genetic exchange in Salmonella*, J. Bacteriol. 64, pp. 679-699.

Acknowledgements
I am indebted to Dr. John Armstrong, who helped turn a garbled transcription of my taped garbled lecture into an understandable and readable draft manuscript. Some of the studies discussed were supported, in part, by grants from the US Environmental Protection Agency and the US National Science Foundation. The opinions expressed herein are not necessarily those of the Agency or the Foundation.

7 DNA and the new organicism

MAE-WAN HO
Biophysicist
Department of Biology, Bioelectrodynamics Laboratory, The Open University
Walton Hall, Milton Keynes, Bucks MK7 6AA, UK

Abstract
The discovery of DNA was the crowning achievement to a century of mechanistic, reductionist biology which firmly established the predominance of the genetic paradigm and of molecular biology. It gave rise to recombinant DNA research, whose very successes, in turn, completely undermined the foundations of the genetic paradigm. The new, post-recombinant DNA genetics signals the final demise of mechanistic biology, and is in accord with a diametrically opposite, organicist perspective emerging in the rest of science. Current practice and commercialization of gene biotechnology, however, are still (mis)guided by the old paradigm of the supposed constancy of DNA in the genome determining the characteristics of organisms in linear, additive fashion. And therein lies the real danger. I review the relevant findings of recombinant DNA research revealing the underlying fluidity of DNA, and the circular, multidirectional network linking all genes. This serves to bring out the hazards of current practices based on outmodified, discredited assumptions. I then present a new organicist view of the organism which properly embeds the fluidity of DNA in an interconnected, intercommunicating whole. This has profound implications for heredity and evolution.

Introduction
The discovery of the DNA double-helix was the climax to a century of mechanistic, reductionist biology – the idea that the whole is the sum of its parts, that cause and effect are simply related, and can be neatly isolated. The discovery ended the quest for the material basis of the units of heredity – the genes – that are supposed to determine the characters of organisms and their offspring, thus firmly establishing the predominance of the genetic paradigm. The subsequent flowering of molecular biology gave rise to the present era of recombinant DNA research and commercial gene biotechnology.

What very few people realize is that the very successes of recombinant DNA research have completely undermined the foundations of the genetic paradigm, so much so that one can legitimately contrast the new, post-recombinant DNA genetics, with the old, pre-recombinant DNA genetics. The new genetics signals the final demise of mechanistic biology, and is consonant with a diametrically opposite, organicist perspective which has been emerging in the rest of science since the beginning of the present century.

Organicism focuses on those fundamental attributes of the organism that the mechanists fail to address: its wholeness and complexity, its dynamism and flexibility, its openness to, and mutual entanglement with the environment, and most of all, its autonomy and active participation in constructing reality (see Ho, 1996a). However, gene biotechnology is practised and sold to the public on the basis of the old, discredited genetic paradigm, and therein lies the real danger.

DNA and the genetic paradigm

A 'paradigm' in science is a comprehensive system of thought and practice developed around a key idea or theory, which can be so pervasive as to spill over into all other disciplines, and to permeate the popular culture. The genetic paradigm is of this nature. It portrays genes as the most fundamental essences of organisms, and supposes that while the environment can be moulded and reshaped, biological nature in the form of genes are fixed and unchanging and can be sorted out from environmental influence. Further, it assumes that the function of each gene is separable from that of every other. It is on such a basis that commercial gene biotechnology is being promoted. We are told that genes 'for' particular characters can be identified, and that transfer of the gene to another organism equals the transfer of the character. We are told that the operation if 'precise' so that nothing else needs to be affected. The Human Genome Project, with strong backing from both state and pharmaceutical industries, promises, not only to identify all genes that make us defective and to cure all ills, but to unveil the very "genetic programme" for making a human being. James Watson, the first Director of the Human Genome Organization (HUGO), set the tone, "We used to think that our fate was in the stars. Now we know, in large measure, our fate is in our genes."

The twin pillars of the genetic paradigm are Darwin's theory of evolution by natural selection and the gene theory of heredity as developed by Mendel and others. Darwin proposed that evolution occurs by natural selection, in which nature effectively 'selects' the fittest in the same way that artificial selection practised by plant and animal breeders ensures that the best, or the most desirable characters are bred and preserved. Darwin's theory lacked a mechanism of heredity and variation. This was supplied by Mendel, who proposed that the (Darwinian) qualities somehow inhere in constant factors (later called genes) determining the organisms' characters, which are passed on to the next generation during reproduction, and variations are generated by rare random mutations in those genes. The combination of Mendelian genetics and Darwinian theory resulted in the 'neo-Darwinian' synthesis from the 1930's onwards. This was an extraordinarily productive period which culminated in the discovery of the DNA double helix and the genetic code, putting the genetic paradigm on a firm molecular footing.

The demise of the genetic paradigm and the new genetics

The major assumptions of the old genetic paradigm, and the findings revealed by the new, post-recombinant DNA genetics are contrasted below.

	The Old Genetics	The New Genetics
1	Genes determine characters in a linear, additive way	Genes function in a complex, non-linear network – the action of each gene ultimately linked to that of every other
2	Genes and genomes are stable and except for rare random mutations, are passed on unchanged to the next generation	Genes and genomes are dynamic and fluid, they can change in the course of development, and as the result of feedback metabolic regulation
3	Genes and genomes cannot be changed directly in response to the environment	Genes and genomes can change directly in response to the environment, these changes being inherited in subsequent generations
4	Genes are passed on vertically, i.e., as the result of interbreeding within the species, each species constituting an isolated 'gene pool'	Genes can also be exchanged horizontally between individuals from the same or different species

All the assumptions of the old genetics are falsified by the findings of recombinant DNA research. The first assumption contradicts all that is know about metabolism and genetics for at least the past 40 years. Interaction between genes has long been recognized, and is referred to as epistasis in genetics textbooks. However, neo-Darwinian theory has tended to ignore the complexities in favour of the supposition that genes can be individually selected by the effects they have on corresponding characteristics of the organisms. Conversely, it is supposed that if a particular gene can be identified in an individual with a given character, then it is possible to predict the character of other individuals possessing the same gene, or at any rate, their propensity towards developing the character. That kind of thinking lies behind the current enthusiasm for 'gene hunting'. The race is on to identify (and patent) genes determining a range of characteristics beyond the classical "single gene" diseases to conditions such as cancer, schizophrenia, homosexuality and alcoholism where nutritional, lifestyle and other socio-economic factors are known to play a large role. The genetic determinism inherent to the old paradigm also gives impetus to patent, for commercial exploitation, gene sequences, organisms, seeds, and cell lines from individuals and indigenous tribespeople. This "patenting of life" offends many people's intuitive sense of being human.

When one takes the scientific findings seriously, one realizes that organisms, like human beings, have tens of thousands of genes in their genome, with multiple variants of each gene. Many of the genes code for the thousands of enzymes catalyzing thousands of metabolic reactions which provide energy for us to do everything that keeps us alive. The metabolic reactions form a complicated, branching network in which the product of one enzyme is processed by one or more other enzymes. Thus, no enzyme (or gene) ever works in isolation. Consequently, the same gene will have different effects from individual to individual, because the other genes (in the 'genetic background') are different. So-called "single-gene defects" – which, in any case, account for less than 2% of all human diseases – are now proving to be very heterogeneous. Many different mutations of the same gene, or of different genes may give the same disease, or not, as the case may be. As a geneticist recently declared, there is "no such thing as a single gene disease" (Mulhill, 1995). Thus, it is impossible, in most cases, to give a reliable prognosis of the disease in individuals who have tested 'positive' in genetic screening. On the contrary, it can cause undue anxiety in those individuals and create genetic discrimination against them, and worse, encourage eugenic practices (Hubbard and Wald, 1993).

Recombinant DNA research has revealed further layers of complexities in cellular and genic processes, many of which destabilize and alter genomes within the lifetime of the organism (Dover and Flavell, 1982; see Ho, 1987, 1995a; Pollard, 1984, 1988; Rennie, 1993; Jablonka and Lamb, 1995); in direct contrast to the static linear conception of the "Central Dogma" of molecular biology that previously held sway. The Central Dogma states that DNA 'makes' RNA in a faithful copying process called transcription. The RNA then 'makes' a protein by a process of decoding called translation. There is supposed to be strictly a one-way "information flow" from DNA to RNA to protein, with no reverse information flow possible (Fig. 1a). In other words, proteins cannot determine or alter the transcribed message in RNA, and RNA cannot determine or alter the genetic message in DNA. In reality, such reverse information flow not only occurs in a wide variety of forms, it is a necessary part of how genes function within a metabolic-epigenetic supernetwork (Fig. 1b).

The 'fluid genome' of the new genetics and hazards of gene biotechnology

A complicated web of feed-forward and feedback reactions are involved in the expression of each gene, or to make a single protein. Numerous proteins are mobilized in the transcription of every gene.

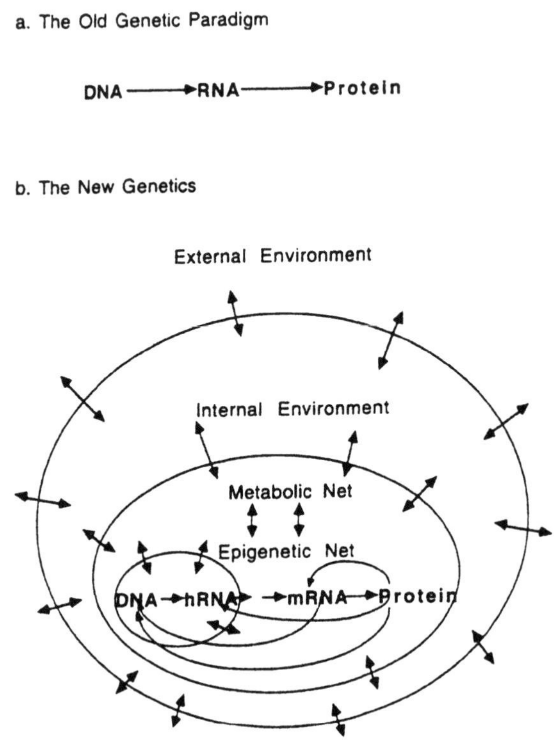

Figure 1. Genetics old and new

Genes, especially of 'higher' organisms, are found to exist in scattered bits, and the bits must be correctly joined together to make the 'messenger' RNA. Other proteins chop and change, even edit and recode in a regulatory or epigenetic network before the final product takes its place in the cell. It becomes increasingly difficult to define and delimit a gene, as the metabolic-epigenetic supernetwork ultimately connects the expression of each gene with that of every other. "Causation" or "information flow" is circular and multidirectional. It is thus impossible to design "magic bullets" by genetic engineering or any other means which would target only one reaction without the effect propagating throughout the system.

So-called "side effects" of pharmaceutical products are familiar enough. The most publicised case in the west was thalidomide. But the problem is widespread in a less dramatic form. According to the 1990 Harvard Medical Practice Study, up to 2 million people are hospitalized in the United States each year as the result of iatrogenic diseases, or diseases caused by prescription drugs, and 180,000 die from them. These are among the visible costs of reductionist biology underpinning mainstream biomedical practice.

The situation will be exacerbated by gene biotechnology, which aims to mass-produce and create new markets for new and old drugs. There will be even less attention paid to health, disease prevention, or to other illnesses of the very young and the old which do not generate profits for the biotechnology industries.

The complexity of the genetic network is also why organisms will tend to change in non-linear, unpredictable ways when even a single new gene is introduced. Unexpected allergenic and toxic products are among the problems of novel foods from gene biotechnology (see Ho and Tappeser, 1996). Many unintentionally sick transgenic animals are created as byproducts of the technology and should raise serious concerns over animal welfare.

The costs of reductionist biology to agriculture and biodiversity are already well-known in the form of the "Green Revolution" (Shiva, 1991). Gene biotechnology will make things worse, for it carries new dangers.

The genome, embedded within the epigenetic-metabolic supernetwork, is far from stable or insulated from environmental exigencies. A large number of processes appear to be designed to alter genomes during the lifetime of all organism, so much so that molecular geneticists have been inspired to coin the descriptive phrase, "the fluid genome". Genes can be marked by chemical modifications, base sequences can mutate, stretches of DNA can be inserted, deleted, or amplified thousands, and tens of thousands of times. The sequences can be rearranged or recombined with other sequences, genes can jump from one site to another in the genome, and some genes can convert other genes to their own DNA sequences. These processes keep genomes in a constant state of flux in evolutionary time. Genes are found to have jumped between species that do not interbreed.

Such horizontal gene transfers, or transfers by infection, were relatively rare events in our evolutionary past. Recently, however, it has been found to be responsible for the rapid spread of antibiotic resistance in bacteria as well as in creating new pathogens. Ho and Tappeser (1996) have focused on the inherent dangers of gene biotechnology in this regard. Genetic engineering depends on the ability to transfer genes between distant species that have extremely low probability of exchanging genes in nature, as for example, transferring human genes into salmon. In order to do that, they rely on specially modified viruses and other genetic parasites as vectors or infectious agents to carry and smuggle genes into the cells that would otherwise reject them. While natural viruses and related genetic vectors are host-specific, generally respecting species boundaries, those constructed by genetic engineers for transferring genes are designed to overcome species barriers and specific cellular mechanisms that break down or inactivate foreign DNA, and are hence potentially much more infectious. Jumping genes, viruses and vectors for gene transfers are all related, and can help one another jump or mobilize, mutate, exchange parts to generate pathogens that can infect each other's hosts. Recent evidence suggests that genes introduced into the environment by transgenic organisms can potentially spread throughout the biosphere via the teeming microbial populations in all terrestrial and aquatic environments (Ho and Tappeser, 1996). These microbial populations form long term reservoirs for the genes to multiply and recombine, and from which they can spread to all species. Again, current practices in making transgenic plants and animals largely ignore the mobility and mutability of genes. The old paradigm gives impetus to create and sell transgenic organisms on the assumption that the transferred genes and the characters they determine will stay in the organisms and remain unchanged ever after. In reality, transgene instability is a recognized problem in both transgenic farm animals and plants (see Ho and Tappeser, 1996 for reference). Their long term commercial viability is yet to be proven. Meanwhile, the hazards from horizontal gene transfer are largely ignored by both the industry and the regulating bodies.

There is much current effort devoted to understanding the nature of biodiversity – how it is that interdependent species in an ecological community nevertheless maintain their autonomy for millions if not hundreds of millions of years. All the signs are that species integrity and the stability of the biodiverse community are intimately linked by a reciprocity of checks and balances, of competitive and mutualistic relationships (Pimm, 1991; DeAngelis, 1991). When sufficiently perturbed, as by the released gene transfer vectors that break down species integrity and species barriers, and especially if current intensive farming and other destructive practices are allowed to continue, such complex non-linear ecological systems may collapse in catastrophic manner (Saunders, 1994).

Most provocative of all, there is now abundant evidence of (previously forbidden) reverse information flow in the genomes of all higher organisms, i.e., information flow from the environment back to the genes. This comes in many forms. Predictable and repeatable genetic changes have been found to occur simultaneously and uniformly in all the cells of the growing meristem in plants exposed to different fertilizers. Plants exposed to herbicides, insects to insecticides and cultured cells to drugs, similarly, are all capable of changing their genomes repeatably by mutations or gene amplifications that render them resistant to the noxious agent, which is why resistance evolves so rapidly, even without the help of transferred genes introducing the resistance. And so long as high levels of herbicides are used with the herbicide-resistant transgenic plants, or high levels of insecticides are expressed by the insect-resistant crops, then some weeds and some insects will be bound to evolve the appropriate resistance, rendering the transgenic plants useless. This has already occurred with plants engineered to be resistant to insect pests with the Bt toxin gene (see Ho and Tappeser, 1996). More and more aggressive use of transgenic plants producing multiple toxins will result in the evolution of multiple resistances in insect pests, with disastrous consequences on agriculture as well as on biodiversity. The advocates for a continual escalation of chemical warfare with nature also see the world in terms of neo-Darwinian natural selection, of ruthless competition in the market place and interspecific warfare in nature.

As a final blow to the genetic paradigm, starving bacteria and yeast cells are now known to respond directly to the presence of (initially) non-metabolizable substrates by mutating the genes required to use the substrate. The genetic responses are so specific that they are referred to as "directed mutations".

In summary, genes function in an extremely complex network, such that ultimately, the expression of each gene depends on that of every other. That is why an organism will tend to change in non-linear, unpredictable ways even when a single gene is introduced. Furthermore, the genome itself is dynamic and fluid, and engages in feedback interrelationships with the cellular and ecological environment, so that changes can propagate from the environment to give repeatable alterations in the genome. Conversely, as demonstrated in current transgenic experiments, introducing a single exotic gene into an organism can impact on the ecological environment. A common soil bacterium, Klebsiella planticola, engineered to produce ethanol from crop waste, was found to drastically inhibited the growth of wheat seedlings. Similarly, the release of transgenic plants with built-in insecticide has led to the rapid evolution of resistance among major insect pests (see Ho and Tappeser, 1996 for references).

The genetic paradigm has collapsed under the weight of its own momentum in the burgeoning new genetics. The genes are far from being the constant essences of organisms, whose effects can be neatly separated from one another or from the environment. There is furthermore, no constant genetic programme or blueprint for making the organism, for the genome can also change even as the organism is developing. How should we see heredity in the light of the new genetics? If the genome itself is dynamic and fluid, where does heredity reside? And what are the implications for evolution? These questions cannot be answered within the mechanistic framework.

The new genetics spells the end of mechanistic biology and compels us to a radically different perspective of organic wholeness and complexity emerging in many areas of contemporary research in the west (see for example, Lovelock, 1979; Ho, 1993; Goodwin, 1994; Laszlo, 1995; Kauffman, 1995; Saunders, 1997). I have developed an organicist approach that is consonant with my experience of nature's unity as well as with the frontiers of contemporary western science (see Ho, 1993; 1994b, 1996a). I shall describe it briefly in the rest of the paper.

Organicism old and new

Organicism was born at the turn of the present century, as Newton's mechanical universe of absolute space and time became fragmented, by Einstein's general relativity, into a multitude of contingent, observer-dependent space-time frames; and quantum theory reveals the separate solidity of objects to be an illusion. Instead, the universe is filled with mutually entangled quantum entities which evolve like organisms, This represents such a decisive break with the mechanistic framework that even today, the implications are not yet fully appreciated. I contrast the mechanical and the organicist universe below.

	Mechanistic Universe	Organicist Universe
1	Static and deterministic	Dynamic and creative
2	Absolute, separate space and time	Contingent, observer(process)-dependent space-times, space and time inseparable
3	Space and time both linear, homogeneous and local	Space-time non-linear, heterogeneous and non-local
4	Linear, isolatable cause and effect	Causes and effects non-linear or circular, and not isolatable
5	Simple location of separate mechanical objects	Delocalized, mutually entangled entities or organisms
6	Cartesian divide between observer and observed	Observer and observed mutually entangled
7	Passive, non-participatory objects buffetted by external forces	Active, participatory and intercommunicating, volitional agents

Organicist philosophy began with French philosopher, Henri Bergson (1916), defender of the romantic tradition – that goes all the way back to Goethe – who rejected the mismatch between our aesthetic, subjective experience of reality and its description given in mechanistic science. Bergson exposed the fundamental irrationality of mechanistic science, especially with regard to its separation of space and time, and its denial of free will. Another influential figure was English mathematician-philosopher, Alfred North Whitehead (1925), who saw physics itself, and all of nature, as unintelligible, without a thorough-going theory of the organism that participates in knowing and in constructing reality. To Whitehead, all entities, from electrons to galaxies, are organisms to varying degrees, and more importantly, mutually entangled with one another. This is a very deep ecological principle – the continued existence of each depends ultimately on the state of every other – which is consistent with contemporary versions of the quantum theory.

Organicist philosophy was embraced by a remarkable, multidisciplinary group, the Theoretical Biology Club, in Cambridge University in the 1930's and 1940's. It included Joseph Needham, distinguished biochemist and embryologist, later to be renowned for his work on the history of Chinese Science, Dorothy Needham, muscle physiologist, the geneticist/developmental biologist, Conrad Waddington, mathematician Dorothy Wrinch, philosopher J.H. Woodger, crystallographer J.D. Bernal, and solid-state physicist Neville Mott. They acknowledged the full complexity of living organization as something to be explained and understood, with the help of philosophy, physics, chemistry, biology and mathematics, as those sciences advance, and in the spirit of free enquiry, leaving open whether new concepts or laws may be discovered.

The project of the Theoretical Biology Club was brought to a premature end when they failed to obtain funding from the Rockefeller Foundation. Organicism has not survived as such, but, as I note elsewhere (Ho, 1993; 1996a), physics itself has become more and more 'organic' with increasing interest in non-equilibrium, non-linear phenomena such as high temperature superconductivity, quantum coherence and non-local quantum superposition of states.

In a way, the whole of science is now tinged with organicist philosophy. 'Consciousness' is currently high on the scientific agenda as decades of intensive investigations in brain sciences have also seriously eroded the mechanistic framework (see Ho, 1996a). This organicist revolution has even spread to economics (see Hodgson, 1993; Omerod, 1994; Ho, 1996d), a discipline rooted in direct analogy to mechanistic physics, where its failures have been all too obvious recently in the major economies in Europe.

It would be an exaggeration to say that there is a conscious movement towards organicism. Nevertheless, it is a Zeitgeist which I personally embrace, and propose to call 'the new organicism' in honour of those who have come before and inspired us, and at the same time, to emphasize the greatly enlarged scope of the endeavour today.

The new organicism, like the old, is dedicated to the study of the organic whole, and does not recognize any discipline boundaries. It is to be found between all disciplines. It is an unfragmented knowledge system by which one lives, as peoples the world over live by their knowledge. No one else, except the western scientist, has an escape clause that allows them to plead knowledge 'pure' or 'objective', and hence having nothing to do with life. Organicist knowledge is resolutely practical. As with the old organicism, the knowing being participates in knowing, and also in transforming and constructing reality, which is why responsibility lies squarely with the knower. It is time scientists assume responsibility for what they do, instead of hiding behind the claim to 'objectivity'. There is, furthermore, no placing mind outside nature as Descartes has done for the mechanistic western knowledge system. Instead, the knowing being is wholeheartedly within nature: heart and mind, intellect and feeling. Organicist knowledge is therefore, not only practical, but also aesthetic and spiritual. In those respects, its affinities are with the participatory knowledge systems of traditional indigenous cultures all over the world.

An organicist theory of the organism

In 1985, I was an evolutionist who has ventured into molecular genetics in order to understand precisely those feedback interactions that can alter genomes in the course of development, which would have significant implications for evolution. A chance meeting with quantum biophysicists Fritz Popp and Emilio DelGuidice at a conference (see Kilmister, 1986) inspired me to study their work and the work of Herbert Fröhlich (1968; 1980). That led to a series of novel experimental investigations into living organization. In the process, I also learned enough of thermodynamics and quantum physics from my patient friends and collaborators to realize that powerful though these conceptual tools are, they remain unable to account for the organism because they too, are based on mechanistic notions of space and time.

I have since developed a tentative theory of the organism within the organicist framework (Ho, 1993, 1994b, c; 1995b, c; 1996a, c, e), showing how the organism does require extensions to the existing 'laws' of thermodynamics and quantum physics. I put 'laws' in quotation marks in order to emphasize that they are not laid down once and for all, but are tools for helping us think, and to be transcended if necessary. It is in this spirit that I present my theory of the organism.

I am by no means alone in thinking that organisms transcend the existing laws of physics. Lord Kelvin, co-inventor of the Second Law of Thermodynamic, specifically excluded organisms from its dominion, so impressed was he at how organisms are able to mobilize energy at will. Likewise, Schrödinger was baffled by the organisms' ability to develop and evolve towards ever increasing organization, seemingly in opposition to the Second Law, which predicts that systems should evolve towards a state of disorganized homogeneity, or thermodynamic equilibrium (see Ho, 1993, 1996a). Schrödinger knew of course that organisms, being open to the environment, can become increasingly organized because of the input of energy, and so there is formally no contradiction of the Second Law. Nevertheless, how the organism can develop, maintain and reproduce its organization from the energy flow is still a mystery.

The idea that open systems can "self-organize" under energy flow is evoked by examples such as the Bonard convection cells that arise in a pan of water heated uniformly from below. At a critical temperature difference between the top and the bottom, a phase transition occurs: bulk flow begins as the lighter, warm water rises from the bottom and the denser, cool water sinks. The whole pan eventually settles down to a regular honeycomb array of flow cells. Before phase transition, all the molecules move randomly with respect to one another. However, at a critical rate of energy supply, the system self-organizes into global dynamic order, in which all of the astronomical numbers of molecules are moving in formation, as though choreographed to do so.

Another physical metaphor for the living system is the laser, where energy is pumped into a cavity containing atoms capable of emitting light. At low levels of pumping, the atoms emit randomly, as in an ordinary lamp. As the pumping rate is increased, a threshold is reached at which all the atoms oscillate together in phase, and send out a giant light track that is a million times as long as that emitted by individual atoms.

These examples tell us that energy input and dynamic order are intimately linked. However, energy flow into the system is of no consequence unless the system can capture and store the energy, so that the energy can circulate and do work before it is dissipated. I have shown how the organism store energy in nested, coupled hierarchies of activities spanning a gamut of space-times from global to local, fast to slow, in such a way that the energy can be mobilized to do work most rapidly and efficiently, that is, with a minimum of dissipation.

The consequences of energy storage for the living system are profound. First and foremost, it frees the organism from the immediate constraints of both the first and second law of thermodynamics, as there is always energy available for whatever it wishes to do, and it becomes, thereby, an autonomous being.

Second, on account of the energy stored, the system is extremely sensitive to specific, weak signals. Sensitivity is indeed a hallmark of living organisms. For example, the eye can detect single photons falling on the light sensitive cells in the retina, which then sends out an electrical signal containing a million times as much energy as the photon it receives. Similarly, a few molecules of specific pheromones in the air is enough to draw males insects to their mates which may be miles away. This exquisite sensitivity is characteristic of all parts of the body because of the energy stored, so that weak signals originating anywhere within or outside the body will propagate throughout the system and become amplified, often into macroscopic action.

Third, energy storage frees the organism from mechanistic constraints. No part of it has to be pushed or pulled into action, for it is vibrant with energy all over. It does not function in mechanical hierarchies of controller versus the controlled, as, in the ideal, healthy system, each part of the system is as much in control as it is sensitive and responsive. So intercommunication is the key to living organization. An organic whole is an interconnected, intercommunicating whole that always has up-to-date knowledge of itself. That is the basis of 'the wisdom of the body' that generations of physiologists have called attention to.

Intercommunication can proceed very rapidly due to the liquid crystalline nature of organisms. Liquid crystals are states of matter in between solid crystals and liquids; hence they combine order with flexibility and responsiveness, and are especially suited for making organisms. Joseph Needham (1936) and others have already predicted that living organisms are liquid crystalline, although they did not have the tools to prove it. The recent discovery of a novel non-invasive optical technique in my laboratory enables us to see living organisms precisely as liquid crystalline.

I now show a video sequence recorded in real time. This is a live Drosophila larva about to hatch, the optical technique enables one to see the whole organism effectively down to the molecules that make up its tissues (see Ho et al, 1996). Brilliant interference colours are generated by polarized light passing through liquid crystalline regimes in the tissues. The principles involved are the same as those in identifying mineral crystals in geology. The larva is obviously alive: waves of muscle contraction are sweeping over its body, so one can infer that all the molecules in the tissues must be moving about busily transforming energy, so how is it possible for them to have a crystalline order? It is because the molecules are moving together in a highly correlated or coherent way. As visible light vibrates much faster than the molecules can move, the tissues will appear indistinguishable from static crystals to the light passing through so long as the movements of the constituent molecules are sufficiently coherent. With this imaging technique, one can see that the activities of the organism are fully coordinated in a continuum from the macroscopic to the molecular, and that is what the wholeness, or the coherence of the organism entails. In effect, all the parts or levels or the organism are constantly intercommunicating, mutually adjusting to one another and yet each is carrying on autonomously doing its own thing. This is so for the fruitfly larva, as for the brine shrimp and the Daphnia and every other organism we have looked at. Elsewhere (Ho, 1993; 1996c, e), I have argued in detail that quantum coherence is the basis of the wholeness of the organism – a special state of coherence that maximizes both local freedom and global cohesion. Such a concept of coherence has considerable implications for social organization (see Ho, 1996f).

This imaging technique also demonstrates how observer and observed are mutually entangled. The colours are generated in the organisms when we set up our observation in a certain way, yet they depend on the physiological and developmental state of the organisms. It is no wonder that as our methods of observations are mechanistic, we shall know the organism as a mechanistic being. Conversely, only as our methods become more and more sensitive and minimally 'invasive' do we finally know the organism in its own right as an authentic organic whole. The knower neither imposes her arbitrary will on nature, nor becomes a passive recipient of sensory information from nature. Instead, there is always a unique entanglement of the knower and the known, much as Goethe had envisaged. Perception is active, and in order to know a living being, it may be said perceiver and perceived are both active. It too, is an act of intercommunication of the same form as that which underlies living organization. In the organicist perspective, therefore, the organism is always sustained by intercommunication with its ecological environment of all other organisms. It is sheer folly to suppose that we can change one species or introduce a new gene into a species without affecting anything else. The organicist attitude to nature is no longer one of mutual warfare but an ecological reciprocity of checks and balances, of gives and takes, precisely the way that our bodies work so efficiently as organisms.

DNA and the new organicism

The organicist perspective gives us a different understanding of DNA which is consonant with the findings of the new genetics. DNA is fluid and dynamic as much as a "gene" is, in many ways, delocalized over the entire genome, each gene being entangled with all other genes. It is clear that DNA cannot be seen as the 'unmoved mover', or central controlling agency directing the development of the organism in a mechanistic hierarchy of controller genes and the genes controlled, for causation is circular and multidirectional. Nor can we suppose that it is the constancy of DNA which is responsible for heredity. As we have seen, there is no constant genetic programme directing the one-way flow of information. So-called control and heredity does not reside solely in the DNA of the genome; instead, it is distributed and delocalized over the whole organism and its environment, all parts of which are interconnected, constantly intercommunicating and adjusting to one another over the whole gamut of space-times that ultimately straddle generations.

In the first instance, heredity resides in an epigenetic cellular and physiological state – a dynamic equilibrium of interlinked genic, cellular and physiological processes constituting a "memory" that is, in reality, always projected towards the future to influence how the system develops and responds (see Ho, 1993). As organisms engage their environment in successive rounds of mutual feedback interrelationships, they transform and maintain their environments which also influence their own future developments, and are further passed on to subsequent generations as home ranges, cultural traditions and artefacts to set the parameters for the development of future generations. It is the entire complex of mutual dynamic interrelationships between organism and environment that gives rise to the stability and repeatability of the developmental process, which we recognize as heredity. The fluidity of DNA and the genome is necessary to the dynamic stability of the system, for genes also adjust as appropriate to the intercommunicating whole.

What implications are there for evolution? Just as interaction and selection cannot be separated, so neither can variation (or mutation) and selection, for the 'selective' regime may itself cause specific variations or 'adaptive' mutations. The organism experiences its environment in one continuous nested process, adjusting and changing, leaving imprints in its epigenetic system, its genome as well as the environment, all of which are passed on to subsequent generations. Thus, there is no separation between individual development and the evolution of future generations. Past, present and future are entangled in nested organic processes. In that respect, our fate is neither written in the stars nor in our genes, for we are active participants of the evolutionary drama. In a very real sense, we never cease to write and overwrite our evolutionary history. DNA may be seen as a specific text in which organisms record their evolutionary experiences. It is a tissue of contingent circumstances and volitional constructions which influences and supports the future survival of the species. It is paramount that we leave a legacy of wisdom to sustain future generations, rather than a record of our folly that will condemn them to extinction.

References

Bergson, H. (1916) *Time and Free Will. An Essay on the Immediate Data of Consciousness* (F.L. Pogson, trans.), George Allen & Unwin, Ltd., New York.

DeAngelis, D.L. (1992) *Dynamics of Nutrient Cycling and Food Webs,* Chapman and Hall, London.

Dover, G. A. & Flavell, Ed. (1982) *Genome Evolution,* Academic Press, London, p. 382.

Fröhlich, H. (1968). Long range coherence and energy storage in biological systems, *Int. J. Quant. Chem.* 2, pp. 641-649.

Fröhlich, H. (1980) The biological effects of microwaves and related questions, *Adv. Electronics and Electron. Phys.* 53, pp. 85-152.

Goodwin, B.C. (1994) *How the Leopard Changed Its Spots, The Evolution of Complexity,* Weidenfeld and Nicolson, London.

Ho, M.W. (1987) Evolution by process, not by consequence: implications of the new molecular genetics for development and evolution, *Int. J. comp. Psychol.* 1, pp. 3-27.

Ho, M.W. (1993) *The Rainbow and the Worm: The Physics of Organisms,* World Scientific, Singapore.

Ho, M.W. (1994a) What is (Schrödinger's) negentropy?, *Modern Trends in BioThermoKinetics* 3, pp. 50-61.

Ho, M.W. (1994b) Towards an indigenous western science: causality in the universe of coherent space-time structures, in W. Hartman and J. Clark (eds.), *New Metaphysical Foundations of Modern Science,* Institute of Noetic Sciences, Sausalito, pp. 179-213.

Ho, M.W. (1995a) Unravelling gene biotechnology, *Soundings* 1, pp. 77-98.

Ho, M.W. ed. (1995b) *Bioenergetics, S327 Living Processes, An Open University Third Level Science Course,* Open University Press, Milton Keynes.

Ho, M.W. (1995c) Bioenergetics and the Coherence of Organisms, *Neural Network World* 5, pp. 733-750.

Ho, M.W. (1996a) The biology of free will, *J. Consciousness Studies,* pp. 231-244.

Ho, M.W. (1996b) Why Lamarck won't go away, *Ann. Human Genetics* 60, pp. 81-84.

Ho, M.W. (1996c, in press) Bioenergetics and biocommunication, in R. Paton (ed.), *IPCAT 95 Proceedings,* World Scientific.

Ho, M.W. (1996d, in press) On the nature of sustainable systems, *World Futures.*

Ho, M.W. (1996e, submitted) *Towards a theory of the organism.*

Ho, M.W. (1996f). Natural being and coherent society, in P. Bunyard (ed.), *Gaia in Action, Science of the Living Earth,* Floris Books, Edinburgh, pp. 286-307.

Ho, M.W., Haffegee, J., Newton, R., Zhou, Y.M., Bolton, J.S. and Ross, S. (1996, in press). Organisms are polyphasic liquid crystals, *Bioelectrochemistry and Bioenergetics.*

Ho, M.W. and Tappeser, B. (1996) Transgenic transgression of species integrity and species boundaries, in K. Mulongoy (ed.), *Proceedings of Workshop on Transboundary Movement of Living Modified Organisms Resulting from Modern Biotechnology: Issues and Opportunities for Policy-makers, Aarhus, Denmark, 19-29 July,* Swiss Academy of the Environment.

Hodgson, D.M. (1993) *Economics and Evolution: Bringing Back Life into Economics*, Polity Press, Cambridge.

Hubbard, R. and Wald, E. (1993) *Exploding the Gene Myth*, Beacon Press, Boston.

Jablonka, E. & Lamb, M. (1995) *Epigenetic Inheritance and Evolution. The Lamarckian Dimension*, Oxford University Press, Oxford, p. 301.

Kauffman, S. (1995) *At Home in the Universe*, Oxford University Press, New York.

Kilmister, C.W. (1986) *Disequilibrium and Self-Organization*, Reidel, Dordrecht.

Laszlo, E. (1995) *The Interconnected Universe – Conceptual Foundations of Transdisciplinary Unified Theory*, World Scientific, Singapore.

Mulvihill, J.J. (1995) Craniofacial syndromes: no such thing as a single gene disease, *Nature Genetics* **9**, pp. 101-103.

Needham, J. (1936) *Order and Life*, MIT Press, Cambridge, Mass.

Omerod, P. (1994) *The Death of Economics*, London/Boston Faber & Faber, London.

Pimm, S.L. (1991) *Balance of Nature – Ecological Issues in the Conservation of Species and Communities*, The University of Chicago University Press, Chicago.

Pollard, J. W. (1984) Is Weismann's barrier absolute? in M.W. Ho and M.W. Ho & P.T. Saunders (eds.) *Beyond neo-Darwinism: Introduction to the New Evolutionary Paradigm*, Academic Press, London, pp. 291-315.

Pollard, J. W. (1988) The fluid genome and evolution. pp. 63-84 in M.W. Ho & S.W. Fox (eds.) *Evolutionary Processes and Metaphors*, Wiley, London, pp. 63-84.

Rennie, J. (1993) DNA's new twists. *Scientific American*, March, pp. 88-96.

Saunders, P.T. (1994) Evolution without natural selection: Further implications of the daisyworld parable, *J. theor. Biol.*, **166**, pp. 365-373.

Saunders, P.T. (1997, to appear) *Evolving the Probable*.

Shiva, V. (1991) *The Violence of the Green Revolution*, Third World Network, Penang, Malaysia.

Whitehead, A.N. (1925) *Science and the Modern World*, Penguin Books, Harmondsworth.

8 DNA at the edge of contextual biology

JOHANNES WIRZ
Molecular biologist
Forschungsinstitut am Goetheanum, Naturwissenschaftliche Sektion
Hügelweg 59, CH-4143 Dornach

Abstract
The rapid progress in the identification of genes and their structural and functional properties has led to a tremendous growth of knowledge of molecular and mechanistic aspects of life processes. As a consequence, the gene concept has undergone a major revolution. Genes are no longer the cause of development and form in living things. They only provide necessary, but not sufficient conditions. A true understanding of life and its processes makes a contextual approach mandatory. It integrates genes, phenotypic traits of organisms and environmental qualities. This approach is the conceptual synthesis of all the observable aspects of what is called a "living being". Contextual biology exhibits consequences in all biological sciences. As an example, the genetics of adaptive mutations are described and a transition from gene-centred to organism-centered inheritance is proposed.

Introduction

The basic assumption of molecular biologists is that life can be described and understood in terms of a genetic program, the regulation of gene expression and the interaction of gene products. In the present paper I would like to challenge this view. It is not the knowledge of molecular processes and cellular functions that results in a concise concept of life, but our recognition of life and life processes that allows for a full comprehension of DNA, i.e. gene function. Thus, life is *not* a consequence of but the *cause* for the evolution of the molecular and cellular machinery of organisms. This reversed view has important consequences with respect to the paradigm of biology.

A thought experiment helps to introduce this reversal. Consider an autoradiogram of a sequencing gel (Figure 1.). Without any knowledge of technical and experimental details, the DNA-sequence can be deduced by determining the order of the black bars in the four rows. If these rows are assigned by T, C, G, A, respectively, the building blocks

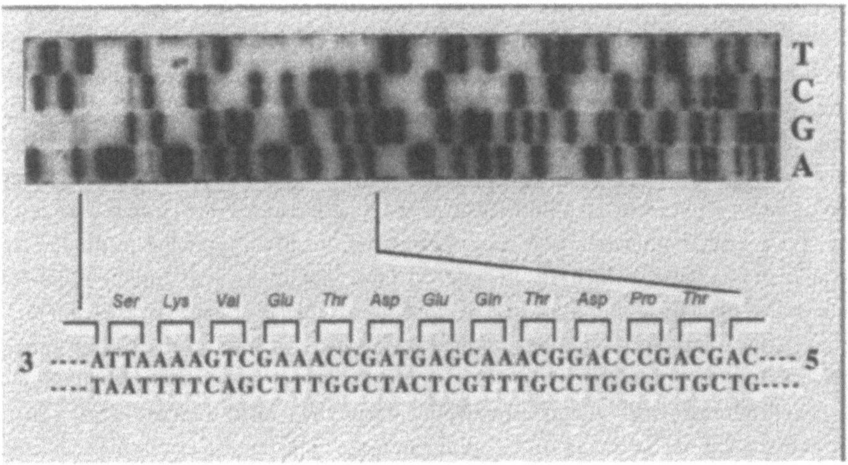

Figure 1. Autoradiogram of a sequencing gel.

Part of the nucleotide sequence is shown. On top of it, the amino acid sequence for a selected reading frame has been deduced.

of DNA shared by all organisms, the gene sequence can easily be written down. This single stranded sequence can be completed. Six possible ways to translate this sequence into the amino acid sequence of a protein are possible.

Interestingly enough, the amino acid sequence or the so called primary structure of a protein is not a sufficient piece of information to elucidate the function of the putative protein. Moreover, it does not exhibit any traces of the organism which it has been derived from. Thus, when the DNA sequence is ready for further manipulation and engineering, it has lost all the organismically relevant content.

DNA at the edge

The significance of this example is emphasized by a large number of publications in the scientific press. Typically, the aim of many genome projects is to determine the entire genomic sequences in a variety of organisms from viruses and microbes to man. The proportion of putative proteins with unknown function deduced from DNA sequences varies from 20% to 80% depending on the organism investigated. For example, out of some 88,000 sequences only some 10,000 could be related to known genes by means of sequence homologies with previously identified genes (Adams et al. 1995). If genes really are the cause for biological processes, we are left with the paradoxical situation that there is a substantial number of "causes" without known "effects".

This DNA paradox is frequently encountered with transgenic organisms, as well. Despite the description in the popular press of giant mice, sex reversal of mice etc., the scientific literature shows the results to be more ambiguous. For example, the sex reversal of mice announced as "Making a male mouse" was only partly successful. Not all the female mice carrying a testis-determining gene became male (Koopman et al. 1991). As Craig Holdrege (1996) has stated feminist scientists could have claimed that "female mice resist attempt at male domination". Or, mice engineered to carry a homolog to the human retinoblastoma gene, which in its mutant form results in eye tumours, showed unexpected phenotypes but in no case retinoblastomas (Lee et al. 1992, Jacks et al. 1992). The Lesch-Nyhan disease in mouse created by targeted mutation of the corresponding gene proved to be asymptomatic (Davies 1992). And mice manipulated to mimic Gaucher's disease died within twentyfour hours after birth (Davies 1992). Such phenomena are not restricted to transgenic animals, only. Similar cases of unexpected results have been reported for bacteria and plants.

Obviously, the presence of the DNA sequence for a given gene is not per se sufficient to direct the formation of characters attributed to it. The context into which the gene is introduced, be it the position on the chromosome, the tissue where it is expressed, the living being as such, or environmental conditions, is an essential factor for determining a gene's effect.

The paradox can be extended to human genetics. One of the classical inheritable diseases, sickle cell anaemia, can be traced back to a single nucleotide change in the haemoglobin gene. However there are many clinical symptoms from early childhood mortality to hardly recognisable symptoms at the age of fifty (Lander and Schork 1994). Cystic fibrosis, a seemingly simple monogenic disease, exhibits a highly complex situation on the molecular level. Some 400 mutations are known with very diverse phenotypes (Alper 1996). The same nucleotide change(s) may lead to different symptoms in different patients. Finally, mutations in the tyrosine kinase gene RET have been attributed to four very different syndromes (van Heyningen 1994).

Common to all examples is the fact that an integral understanding of gene function is not solely dependent on the sequence information. The bottom to top view, i.e. from gene to organismic trait can be deceiving. It has to be complemented by a top to bottom view, from organism to gene, reversing the reductionist approach. Genes and their products mediate processes and therefore represent necessary but not sufficient *conditions* for biological processes.

Overcoming gene-centred biology

These limitations of gene-centred biology have intrigued many scientists for quite some time. New approaches have been developed to enlarge and deepen our view of organisms and life. The efforts of the science of complexity should be mentioned, which showed relational order and dynamic interactions between components to be more important than the material, i.e. the chemical composition of life (Ho and Fox 1988, Goodwin 1994, Kauffman 1994, Lewis 1993). However, none of these approaches have appreciated the very particular nature of "relational order" and of "complexity". They have overcome the methodological deterministic materialism related to gene-centred biology but not the ontological one. Thus, whereas it seems clear that *genes* are not causes for biological form and life, *matter* as such has taken the role of causative creation. How does this come about?

There is no doubt that every experimental design, the experimental result and its subsequent interpretation are dependent on a conceptual framework, on a specific and concise *context*. As a pre-theoretical decision (Rehmann-Sutter 1996) which determines a particular scientific approach towards life and life processes, it escapes our attention. As a scaffold which helps to structure knowledge and theory, it fades in the face of hard facts and goes unnoticed. The reason for this has to do with the uniqueness of context. Context as a concept cannot be touched on, not be looked at and even less be extracted and isolated from organisms. Like all concepts, it resists the transformation into an object of the material world. *Context as a conceptual principle is accessible via our intellectual, thinking capacity, only. Therefore it unravels the spiritual nature of life.*

Contextual biology is far from the renewal of the dualistic matter-mind paradigm or the reintroduction of the old theory of vitalism. It recognises that fact that the reality of life and living beings is based on two aspects: All the perceptual qualities *as well as* the conceptual ones. To use a metaphor of Portmann (1973): The play of life can be observed from two different sides. As spectators we may enjoy the exciting story brought on stage by outstanding actors. In this case we deal with the proper *meaning of life*. But we can also admire the technical equipment behind the curtain necessary to put the play on stage. In this case, we deal with all the *functions* that are required for life to appear. Contextual biology acknowledges both of them, the "backstage" function of life as well as the "frontstage" meaning of life.

The theoretical foundation of *contextual biology* has been given in the pre-molecular era of biology (Steiner 1886) on the basis of an unbiased appreciation of both the sense perceptible and the conceptual, i.e. spiritual qualities by which the reality of life is structured. But it is only today that the results of molecular biology themselves make the recognition of context obligatory.

Contextual biology

Contextual biology calls for a clear cut methodology to discriminate the two aspects that relate to all living beings. The first includes *all* sense perceptible and material qualities: DNA, proteins, cells as well as the many abiotic and biotic environmental conditions. The second aspect consists of the principle that unifies and integrates all these material qualities (Holdrege 1996 has given a detailed analysis of context). According to Goethe (1820) it can be called the *idea of the organism* or the *archetype*. Whereas the physical, chemical and biotic factors define the set of the necessary conditions for a living being to appear, the archetype allows for their proper specification. It provides the meaning for the appearances of life. The archetype therefore is the ultimate cause for the manifestation of life.

These considerations have been reformulated in the two irreducible laws of life sciences (Goethe, cited in Steiner 1891): The law of the outer circumstances (Gesetz der äusseren Umstände) and the law of the inner nature (Gesetz der inneren Natur). The former includes all observable, material qualities from genes, molecules and cells, to environmental factors. The latter provides the concept or idea to relate these qualities in a meaningful way. The inner nature of a living being transcends the material world. Contextual biology escapes the danger of reducing life processes and living beings to genes and gene expression. It transforms a gene centered into a truly organism centered biology.

The impacts of contextual biology should not be underestimated. First, they lead to a total integration of all observations, descriptions and interpretations of life processes into a unifying theory. Second, they require a decent methodological reflection of presuppositions and results. Third, they lead to the conclusion of a mutual interdependency of the inner nature and the outer circumstances. As a true holistic approach, contextual biology bridges the mind-matter gap.

Complementary genetics

Let us consider, as an example, the consequences of contextual biology in genetics. If the conclusion of mutual interdependency is correct, the current paradigm has to be complemented. Inheritance is not only driven unidirectionally by genes and DNA, but also by the life history of organisms themselves (for a detailed description see Wirz 1996). Besides the random variation, which is basically due to errors during DNA replication and could be called *gene-centred mutation*, there must exist a second genetic variation that is dependent on the interaction of the organism with its environment, an could be called *organism-centered mutation*. In contrast to *random* variation, where a spontaneous mutational event eventually leads to an altered phenotype, *adaptive* variation is induced by the organism in a directed way, according to a preceding change in the phenotype. Thus, "cause" and "effect" of genetic variation are reversed in these two types of variation.

Directed mutation has been giving very little credit for two reasons. Firstly, it is considered to be "Lamarckian", despite the fact that Darwin himself and some of his successors had integrated this type of variation into their theories. Secondly, after the discovery of spontaneous mutations by Luria and Delbrück (1943) they quickly became a "dogma" in genetics, even more so after the postulation of the "central dogma" (Crick 1970). Strong experimental evidence for spontaneous mutations was given by the molecular analysis of DNA replication.

Despite its apparent success the central dogma has been challenged by a number of scientists (see e.g. Ho and Fox 1988). Waddington, for example, (1959) questioned the random variation theory since certain organismic traits like the forelimbs of gibbon or pangolin required the concerted variation of many characters. They looked as if they were dependent on and therefore the result of the specific behavioural activities of their carriers. Bockemühl (1981) and Holdrege (1986) postulated on the basis of a detailed analysis of the leaf metamorphosis of a garden weed (Senecio vulgaris) during the course of the year that the genetic constitution of the plant, rather than promoting variation, seemed to restrict it. And Cullis (1988) showed that genotypic variation in flax could be brought about via specific environmental conditions.

These examples show that the *interaction between organisms and their surroundings plays a central role in the establishment of the genetic constitution and its variation.* However, these approaches – and variety performed by other scientists – failed to show a directed environmental influence on the genotype or they failed to give an account for the significance of the environmentally induced genetic alteration for the organism.

This situation changed with the publication of directed mutation in E. coli (Cairns et al. 1988). During the subsequent controversial discussion many more examples were presented (for a review see Foster 1993) and serious arguments against it followed (Mittler and Lenski 1993). Today, the phenomenon of *directed mutation*, also called *adaptive mutation* or *selection induced mutation* is established for many microorganisms and catabolic, as well as anabolic processes. Experiments to elucidate its molecular mechanisms have been reported (Hall 1991, 1993, Harris 1994, Galitski and Roth 1995, Radicelli et al. 1995).

In our laboratory we have started to investigate the phenomenon of adaptive mutation in Drosophila. Preliminary results (Wirz unpublished) show that an adaptive genetic response can be observed for both the variation of quantitative traits – by determining the lifespan of the flies on a novel food source – and enhanced reversion of mutations under inducing conditions – reversion of a dominant temperature-sensitive mutation at elevated temperature.

Table 1. summarizes the conditions for spontaneous, i.e. *gene-centered* and adaptive, i.e. *organism-centered* mutations. It turns out in all experiments that the conditions that lead to either types of variation are complementary in all aspects. The former are uniquely dependent on general molecular processes during replication, the latter depend on specific interactions between organism and environment. Spontaneous mutations strongly depend on replication and hence cell proliferation. By contrast, adaptive mutations occur during phases of slow or even arrested growth. Interestingly, the spontaneous mutation paradigm has been developed from studies with exponentially growing bacteria – hence the results reflect the scientists' presuppositions for the experiments.

Table 1

Comparison of the conditions that promote spontaneous and adaptive mutations, respectively.

spontaneous mutations	*adaptive mutations*
independent of life history	dependent of life history
undirected	directed
replication dependent	replication independent
selection after variation	selection induced variation

The occurrence of adaptive mutations does not conflict with those that are spontaneous. However, this type of genetic variation would not have been anticipated by a purely molecular paradigm, which upholds the unidirectional flow of information from genotype to phenotype. Adaptive mutations provide a compelling evidence for a reversed flow of information. They strengthen the notion of continuum between organism and environment and extend it to the hereditary processes. They stress the significance of a complementation of molecular biology and organismic biology. Moreover, adaptive mutations illustrate a necessary consequence of the two laws of contextual biology: *Variation flows both from genes to organisms and from organisms to genes.*

Adaptive mutations unravel in still another way the dichotomy between meaning and function. To understand meaning we have to observe and investigate the living being in its specific environment. To understand function, i.e. to learn how these mutations come about, we need to investigate the molecular processes which mediate them.

Conclusion

The present paper has shown first of all that the paradigm of molecular biology needs complementation. The concept of contextual biology allows for a proper understanding of the possibilities and limitations of the molecular approach and calls for its quantitative, as well as qualitative enlargement. It (re)introduces the concept of the spiritual nature of life and living beings in biology. The genetics of adaptive mutations has been presented as an example for the complementation of the current paradigm of molecular genetics. This example has stressed the necessity to investigate – beside molecular interactions and processes – organismic relations between living beings and their environment.

Contextual biology reintroduces the concepts of *meaning* and *values* in the life sciences. By so doing it provides the foundation of a overall holistic biology, that is not restricted to the totality of qualities and processes in the world outside man, but includes the human being not only on the basis of partner or competitor sustaining natural evolution or abolishing it but also with his most specific humane quality: the potential to grasp and to structure concepts and ideas. In this sense, contextual biology transcends most of the current holistic approaches towards nature, since it recognizes the spiritual quality of life.

References

Adams, M.D. et al. (1995) Initial assessment of human gene diversity and expression patterns based upon 83 million nucleotides of cDNA sequence, *Nature Suppl.* 377, pp. 3-174.

Alper, J. (1996) Genetic complexity in single gene diseases, *BMJ* **312**, pp. 196-197.

Bockemühl, J. (1981) A new perspective on heredity, in J. Bockemühl (ed.), *In Partnership with Nature*, Bio-Dynamic Literature, Wyoming, pp. 31-33.

Cairns, J., Overbaugh, J. and Miller, S. (1988) The origins of mutants, *Nature* 335, pp. 142-145.

Crick, F. (1970) The central dogma of molecular biology, *Nature* 227, pp. 561-563.

Cullis, C.A. (1988) Control of variation in higher plants, in M.-W. Ho and S.W. Fox (eds.), *Evolutionary Processes and Metaphors*, John Wiley & Sons, Chichester etc., pp. 49-62.

Davies, K. (1992) Mulling over mouse models, *Nature* 359, p. 86.

Foster, P. (1993) Adaptive mutations: the use of adversity, *Annu. Rev. Microbiol.* 47, pp. 467-504.

Galitski, T. and Roth, J.R. (1995) Evidence that F plasmid transfer replication underlies apparent adaptive mutation, *Science* **268**, pp. 421-423.

Goethe, J.W. (1820) Erster Entwurf einer allgemeinen Einleitung in die vergleichende Anatomie, ausgehend von der Osteologie, *Zur Morphologie* 1, cited in: Johann Wolfgang Goethe Sämtliche Werke, Buchclub Ex Libris, Zürich (1979).

Goodwin, B. (1994) *How the leopard changed its spots*, Weidenfeld and Nicolson, London.

Hall, B.G. (1991) Spectrum of mutations that occur under selective and non-selective conditions in E. coli, *Genetica* **84**, pp. 73-76.

Harris, R.S., Longerich, S. and Rosenberg, S.M. (1994) Recombination in adaptive mutation, *Science* **264**, pp. 258-260.

Holdrege, C. (1986) Schritte zur Bildung eines lebendigen Vererbungsbegriffes, *Elemente der Naturwissenschaft* **45**, pp. 27-61.

Holdrege, C. (1996) *Genetics and the manipulation of life: the forgotten factor of context*, Lindisfarne Press, Hudson, NY.

Jacks, T., Fazeli, A., Schmitt, E.M., Bronson, R.T., Goodell, M.A. and Weinberg, R.A. (1992) Effects of an Rb mutation in the mouse, *Nature* **359**, pp. 295-299.

Kauffman, S. (1995) *At home in the universe. The search for laws of complexity*, VIKING, Penguin Books Ltd, London.

Koopman, P., Gubbay, J., Vivian, N., Goodfellow, P. and Lovell-Badge, R. (1991) Male development of chromosomally female mice transgenic for Sry, *Nature* **351**, pp. 117-121.

Lander, E.S. and Schorck, N.J. (1994) Genetic dissection of complex traits, *Science* **265**, pp. 2037-2048.

Lee, E.Y.H.P., Chang, C.Y., Hu, N., Wang, Y.C.J., Lai, C.C., Herrup, K. and Lee, W.H. (1992) Mice deficient for Rb are non-viable and show defects in neurogenesis and haematopoiesis, *Nature* **359**, pp. 288-294.

Lewin, R. (1993) *Complexity – life at the edge of chaos*, Macmillan Publishing Company, New York.

Luria, S.E. and Delbrück, M. (1943) Mutations of bacteria from virus sensitivity to virus resistance, *Genetics* **28**, pp. 491-

Mittler, J.E. and Lenski, R.E. (1993) The directed mutation controversy and Neo-Darwinism, *Science* **259**, pp. 188-194.

Portmann, A. (1973) *Biologie und Geist*, Suhrkamp, Frankfurt a.M.

Radicella, J.R., Park, P.U. and Fox, M.S. (1995) Adaptive mutation in Escherichia coli: a role for conjugation, *Science* **268**, pp. 418-420.

Rehmann-Sutter, C. (1996) *Das Leben beschreiben über Handlungszusammenhänge in der Natur*, Königshausen und Neumann, Würzburg.

Steiner, R. (1886) *Grundlinien einer Erkenntnistheorie der Goetheschen Weltanschauung*, Rudolf Steiner Nachlassverwaltung, Dornach (1979); English translation: *Theory of knowledge based on Goethe's world conception*, Anthroposophic Press, New York, (1968).

Steiner, R. (1891) Über den Gewinn unserer Anschauungen von Goethes naturwissenschaftlichen

Arbeiten durch die Publikation des Goethe-Archives, in R. Steiner, *Methodische Grundlagen der Anthroposophie*, Rudolf Steiner-Nachlassverwaltung, Dornach (1961), pp. 265-287.
Van Heyningen, V. (1994) One gene – four syndromes, *Nature* 367, pp. 319-320.
Waddington, C.H. (1959) Evolutionary adaptation, in C.H. Waddington, *The Evolution of an Evolutionist*, University Press, Edinburgh (1975), pp. 36-59.
Wirz, J. (1996) Schritte zur Komplementarität in der Genetik, *Elemente der Naturwissenschaft* 64, pp. 37-52.

Acknowledgements
I wish to thank John Armstrong for his patient and constructive criticism on an earlier version of this paper and David Heaf for comments and final corrections. Georg Iliev's help for the fine-art production of figure 1 is greatly appreciated.

9 The role of genetic disposition in human health and disease – bioethical aspects of DNA testing

HANSJAKOB MÜLLER

Human geneticist

Department of Medical Genetics, University Children's Hospital, Römergasse 8, CH-4005 Basel

Abstract

The aim of the Human Genome Project, an international collaborative effort currently in progress, is to identify and catalogue all human genes and to identify the underlying DNA nucleotide sequences. It provides a basis for determining how genes give rise to human diseases, thus opening up new approaches to their cure and prevention. The efforts to map the human genome present no inherent ethical problems but are eminently worthwhile, particularly since the knowledge they provide will be universally applicable for the benefit of human health. In terms of ethics and human values, it must be ensured that the methods used in gene mapping adhere to the high standards of research ethics and that the knowledge gained will be used in an appropriate manner and only for the benefit of the persons tested. This presents a substantial challenge to the medical geneticist. Interdisciplinary discussions on expectations and misgivings provide him with a unique opportunity to broaden his own understanding of important concepts such as human individuality, family bonds or health and disease.

Introduction

Medicine has entered a new era characterised by a better understanding of the molecular mechanisms underlying health and disease. Gene technology provides us with a deeper insight into the biological processes taking place in our bodies. The actual causes of disease can be determined with ever greater accuracy, thus allowing treatment to be targeted more precisely upon them. In this new era, medicine will tend to concentrate less on treatment and more on prevention.

Predisposition and nature of human disorders

Human diseases may be due to influences of which we still have only a vague understanding such as chance factors or ageing, as well as to defined environmental factors such as malnutrition or microbial infection or to genetic predispositions. Most human diseases result from a combination of such influences. They are multifactorial in origin. This article is concerned with the evaluation of the importance of genetic influences, in the causation of disease.

Predisposing factors range from a single base substitution within the DNA segment representing a particular gene, as in sickle-cell anaemia, to the addition of a complete chromosome as in Down's syndrome (trisomy 21) or the loss of a complete chromosome as in Turner's syndrome (45,X). Although diseases of mainly genetic origin are rare if considered individually, if taken as a group they are numerous and often of a severe character. At least 2% of all live-born babies have genetic disorders (Emergy and Rimoin, 1990). Multifactorial diseases are generally triggered by combined effects of multiple genes at different loci, each of which has a small but additive effect, and also involve environmental factors. Genetic factors can play a substantial role in the causation of many common diseases which manifest themselves only in adulthood, such as coronary heart disease or cancer (Müller, 1993).

The traditional family doctor often used his own intuition to ascertain the presence of a familial predisposition to disease. He personally knew many members of a single family, and thus got to know their individual susceptibilities. With ever greater specialisation and progressive concentration on individual organs or organ systems, doctors increasingly failed to see the characteristics of particular families. They often omitted to consider not only the family, but even the patient himself as an individual. This development led to a crisis in the orthodox school of medicine from which it has not fully recovered even today.

Together with the increasing understanding of molecular genetics and molecular mechanisms that lead to disease, today there is a growing appreciation that the doctor must first of all consider the patient as a person. The same molecular defect in the genome may have completely different effects among affected individuals and families. In medicine, we are always concerned with people and not simply with cases. In this connection we should bear in mind that in our society grandmothers and aunts in particular have a considerable knowledge of inheritable characteristics, notably when it comes to recognising traits shared by an offspring and his antecedents.

A parallel in medical history

Medicine in the form in which it is now taught in our universities and practised would be inconceivable without the investigations into our body's anatomy which began in the Middle Ages. In the 16th century Andreas Vesalius wrote the first comprehensive and correct description of human anatomy (Vesal, 1543). Pathologists subsequently compared healthy and diseased organs. Even today, morphological changes are still the "yardstick" for the diagnosis of diseases and for the evaluation of medical interventions. When the architecture of the body's organs was known, physiologists began to investigate their function. At the beginning of the 17th century the London physician William Harvey (1628) gave his historic concept of the circulation of the blood. Not until this century did pathophysiology truly establish itself as a branch of medicine.

With the Human Genome Project, work began on investigating the anatomy of the human genome (Green and Waterston, 1991; Landner, 1996). This is composed of 50,000 to 100,000 or even more genes, of which only around 10,000 are so far known. In this mapping project, the individual genes are first identified and then located on the chromosomes (the carriers of the DNA in the cell nucleus), after which their biochemical structure is analysed (McKusick and Amberger, 1993). The deciphering of the succession of base pairs in the DNA double helix is known as sequencing. Very little is yet known about gene physiology, i.e. the regulation of gene activity and the interactions of the genes and their products. In view of the present intensity of molecular genetic research, the deciphering of molecular anatomy and physiology is unlikely to take as long as was needed to gain an understanding of conventional anatomy and physiology.

Genetic testing for medical purposes

As a result of the ever-better recognition and understanding of the significance of predisposition as a causative factor for disease, molecular genetic tests are becoming an increasingly common feature of everyday medicine. When assessing their impact on bioethics, it is easily overlooked that other medical genetic tests also allow concrete conclusions on characteristics of the genotype to be drawn (Table 1). The boundary between conventional and molecular genetic tests is more fluid than is often assumed. A precise conclusion on a characteristic of the genotype can often be drawn from clinical observation alone.

Genetic disorders (rarely those of chromosomal origin, but frequently those of monogenetic origin) may follow a familial pattern. If the same or similar symptoms of a genetic disorder occur in blood relatives of a patient, they often have a similar importance for the assessment of the patient's condition. The systematic establishment of a detailed medical family history is one of the best and simplest methods of diagnosing hereditary diseases.

Today the scientific literature is full of reports on the identification and characterisation of genes which, when defective, are responsible for severe human disorders. This can easily be illustrated by the example of neurological diseases. A steadily growing number of slowly progressive neurodegenerative diseases can already be reliably diagnosed and

Table 1

Spectrum of medicogenetic test possibilities

– clinical observation (overall impression)

– clinical examination (e.g. measurement of size of body or parts of body such as skull, back, legs or detection of "café-au-lait" spots for the diagnosis of Recklinghausen's disease

– imaging processes: X-ray, ultrasound, magnetic resonance spectography

– macropathological, histological and cytological examinations

– biochemical analyses of proteins and other molecules

– cytogenetic examinations (number and structure of chromosomes

– molecular genetic analyses (molecular structure and function of the nucleic acids (DNA/RNA)

classified in their early phases by the demonstration of a mutation in the responsible gene (Martin, 1993).

For the medical management of persons who are suffering from genetic disorders or have a genetic of developing such disorders, it is essential to know exactly what one is talking about. The possibility of demonstrating a specific pathogenic mutation therefore represents a decisive advance in medical genetics and molecular medicine.

Prenatal genetic testing

Until the 1950's, couples at risk of having a child with a severe disorder had very limited options in terms of further reproduction. Diagnostic genetic tests on the embryo or foetus were developed in response to their needs. In 1978, DNA isolated from cells shed in the amniotic fluid was first used for prenatal diagnostic identification of the sickle cell defect in an at-risk foetus (Kan and Dosy, 1978).

All cells which make up an organism are practically identical in their DNA content: they contain the complete genome. This means that any nucleated cell can be used for DNA analysis. This situation is ideal for prenatal DNA diagnosis. Sources of embryonic/fetal DNA are chorionic villus cells, fetal cells shed in the amniotic fluid, fetal blood cells or cells obtained by biopsy of specific tissue. Recently, Cheung et al. (1996) developed a new procedure for prenatal diagnosis of monogenetic diseases using fetal cells from the maternal blood. Such cells have long been known to be present in the maternal circulation.

Presymptomatic genetic testing

Disorders caused entirely or to a substantial extent by an abnormality of the genome are not necessarily congenital, i.e. manifest at birth. This category includes forms of many common diseases of adulthood, such as coronary artery disease, high blood pressure, many cancer, diabetes mellitus, rheumatoid arthritis and some psychiatric disorders. Persons whose genes carry an alteration responsible for such developing conditions may have no obvious symptoms until well into early adulthood or even later. However, the genetic defect can be diagnosed by molecular or cytogenetic tests in every stage of life and even before birth. Such investigations are known as presymptomatic or preclinical predictive testing (Müller, 1993). A distinction is sometimes made between presymptomatic testing for clearly inherited diseases with a high penetrance (Huntington's disease) and susceptibility testing for the identification of genetic contributions to common chronic diseases, such as HLA characteristics. Through the latter tests it is possible to identify people who are only at increased risk of developing a disease but who may never suffer from it. In practice, however, such a distinction is quite artificial since the grey zone between the two entities is quite large.

Presymptomatic testing is not so new as is generally assumed in the public discussion. From the beginning, health care professionals tried to recognise and interpret signs in their patients that would allow them to classify a clinical state and/or to provide better medical support in the future. In some individuals presymptomatic diagnosis can be done by clinical examination alone. For instance, benign polyps or congenital hypertrophy of the retinal pigmented epithelium (CHERPE's) indicate an increased risk of colorectal carcinoma, while cataracts are a sign associated with acoustic neuroma (NF2). Also, presymptomatic laboratory tests were already used in the era before gene technology, e.g. creatine phosphokinase levels for Duchenne and Becker's muscular dystrophy or the determination of vanillylmandelic acid (VMA), metanephrines and catecholamines for phaeochromocytoma, or the short calcium infusion test and the pentagastrin test to identify medullary thyroid cancer, both malignancies belonging to the inherited multiple endocrine neoplasia syndrome II = MEN2 (Lips, 1985). The human genome project is steadily increasing the number of diseases accessible to presymptomatic molecular genetic testing.

Searching for mutated genes which predispose to disorders that manifest themselves later in life may have a significant impact on human health, and has the following purposes:

- to detect a serious genetic disorder before the onset of debilitating symptoms, allowing early treatment and/or the prevention of complications
- to identify people at risk of contracting a disease where life style or exposure to given environmental risk factors are important (ecogenetic trait)
- to develop new medical strategies for persons at risk, e.g. chemoprevention or gene therapy
- to make informed life-planning, including reproductive choices (family planning) possible

A rush to test outside the research environment or within family settings has started. However, population screening should be considered only for treatable or preventable conditions of relatively high frequency, and this, only if genetic counselling of probands can be guaranteed.

Genetic screening

Genetic testing can be performed on a single person, on members of a family, on the population as a whole, or only on subsets of the population without regard for a specific risk. The systematic search for persons having a particular genetic characteristic is called "genetic screening". In contrast to medical genetic diagnosis, screening is not usually carried out at the instigation of the test subjects, but more often on the initiative of the test providers.

Screening programmes must meet considerable legal, ethical and psychosocial requirements. All screening programmes should be preceded by education of the populations to be screened. Neonatal screening has become established in numerous countries on the initiative of paediatricians without thorough pre-evaluation. Countless neonates all over the world have already benefited from such screening through dietary measures which have saved them from the serious consequences of certain metabolic diseases (Thalhammer, 1975).

Premarital screening for carrier status for disorders common in a particular community is already realized (Cao, 1989). Heterozygotic carriers of the gene for thalassaemias are common in Mediterranean countries, for instance, and carriers of Tay-Sachs disease are frequently seen among Ashkenazi Jews (Ludman, 1986). By means of these tests, prenatal diagnosis can already be carried out in the first pregnancy of two heterozygotes.

The list of possible genetic screening programmes is long. Today it is being considered if and in whom such tests would be warranted among persons with a frequent predisposition to neoplastic disease such as breast and colorectal cancer (Müller, 1990).

Limits and concerns of genetic testing

The ability to determine an individual's risk for genetic diseases before he or she is born or before the clinical onset of symptoms, or to identify carriers of a mutated gene who suffer no immediate consequences for their own health, raises an array of ethical, psychological and social issues and has also set off a storm of controversy within the medical establishment and with in the public.

The controversy centres upon the still uncertain interpretation of the test result. If a test affirms the presence of a genetic BRCA1 or BRCA2 mutation, a woman with a family history of breast cancer faces an 85 percent risk – not a certainty – of developing the disease. But the risk cannot be estimated with anything like this accuracy in a woman with the same or a similar mutation who has no relatives with the same malignancy. Such uncertainties present a great challenge in the management of women at risk. In many cases not all genes are identified which, if mutated, can lead to a given disease. Here, we are concerned with the still poorly evaluated heterogeneity which underlies many human diseases (Müller, 1991).

In addition to these biological problems a number of technical obstacles must be cleared before such testing becomes widely practised. Despite the steady advances in molecular genetic disease, the devising of reliable tests that will detect all disease related mutations remains a challenge. Genetic tests per se are never perfect and will not allow all mutations to be detected. Even if tests improve in reliability and extend the range of what they can predict, the question of the interpretation of their results will persist. This is overshadowed by what is probably the most difficult question: What is, from the medical viewpoint, normal and abnormal?

Table 2

Reasons for the limited predictive power of genetic tests
– heterogeneity of genes leading to the same clinical phenotype
– limited analytical sensitivity for allelic diversity
– modifiers of the trait responsible for incomplete penetrance or variable expressivity
– phenocopy
– technical limitations

Genetic counselling before offering genetic tests

Genetic diseases can have a profound impact on individuals and families. Problems related to them can affect persons over long periods of their life and even from birth to death, and influence most important decisions they make during their lifetime.

Disease and illness are not identical concepts. There are substantial differences in the way in which patients and their families perceive the symptoms of a disease that can be clearly defined in medical terms. This also applies, in particular, to the evaluation of genetic risks. Risk perception depends on personal attitude, various circumstances relating to risk justification, and socio-cultural factors; it may also change quite quickly both in patients and among the medical profession.

The ambiguities and anxieties that accompany genetic testing can be addressed through proper genetic counselling (Müller, 1996). Therefore, genetic counselling must be an integral part of practically all types of genetic testing.

The key ethical principles of genetic counselling as recognised by the community of medical geneticists are summarised in Table 3. Genetic counselling is a multifaceted process involving topics associated with diagnosis, risk assessment, the explanation of complex medical and genetic facts and the choice of possible options to deal with the genetic burden. Its goals are so wide-ranging that a single approach can hardly encompass the varied needs of people who seek help. There is no single perfect procedure or technique of genetic counselling.

Table 3

Key ethical principles of genetic counselling
Freedom of a person or a couple
– to make autonomous, informed decisions
– to refuse medical action or information (right to know and right not to know)
– to effect complete diagnosis after informed consent
– to obtain assurance of confidentiality

An accurate diagnosis of the underlying predisposition is the cornerstone of meaningful genetic counselling. All means available to modern medicine and genetic testing must be used in order to meet this central requirement.

One important way to ensure autonomy with respect to genetic testing is to provide adequate information on whose basis a person can decide whether or not to undergo testing. Voluntariness is the cornerstone of any genetic testing program. Proper informed consent in medicine generally involves the presentation of information about the risks, benefits and efficacy of the procedures undertaken and suggested by the medical establishment and the alternatives to them.

DNA and human biography

Today, we medical geneticists are faced by unusual challenges. It is our task to give medical assistance to people with genetic health problems and their relatives and to alleviate their fears and anxieties. The assistance we offer must be in line with the latest developments in medicine and modern genetics, i.e. it must have a sound scientific basis and must be as exact as possible. On the other hand the emphasis must be varied according to the individual case, and it is necessary to treat the patient with compassion and understanding and to avoid giving him the feeling he is being pushed one way or another. The choice of language can thus have a powerful effect on the perception of people with genetic problems.

The possibilities and limits relating to the medicotechnical conduct of molecular genetic tests have been mentioned before. The moral and ethical conflicts that can face medical geneticists in dealing with individuals with clear needs and wishes are now becoming increasingly accentuated, since genetic advice and genetic examinations are being sought by more and more people from different cultural groups, some of whom have different values and beliefs from those of our own society.

In the evaluation of our own medicogenetic actions we depend on our own impressions gained more from our experience of individual cases. There is still a widespread lack of systematically conducted studies on the needs of persons seeking advice and on their ability to cope with the advice given. There is a clear need for research in this field.

As a result of gene technology, present-day medicine is undergoing a rapid process of change. This development cannot be stopped. Instead of bewailing the problems associated with modern genetics and molecular medicine, we should try to think ahead and consider how we can avoid the undesirable consequences of molecular genetic diagnostics for all concerned, whatever the purpose for which it is used. This is a challenge not only to the legislator whose task is to protect the interests of the examined persons vis-à-vis insurers and employers. Above all, it is a challenge to us doctors, because we must learn the right way to analyse complex data and to communicate the results that emerge from them. At the same time, however, we must seek new ways to ensure that the newly available information is used for the medical benefit of the persons examined. In this connection conferences like the one on "The Future of DNA" are an excellent opportunity to learn to understand different points of view and to broaden the understanding of concepts such as human individuality, family relationships, health and disease.

References

Cao, A., Rosatelli, C., Galanello, R., Monni, G., Olla, G., Cossu, P., Ristaldi, M.S. (1989) The prevention of thalassemia in Sardinia, *Clinical Genetics* **36**, pp. 277-285.

Emery, A.E.H., Rimoin, D.L. (1990) Nature and incidence of genetic disease, in A.E.H. Emery, D.L. Rimoin (eds.), *Principles and Practice of Medical Genetics*, Churchill Livingston, Edinburgh, pp. 3-6.

Green, E.D. and Waterston, R.H. (1991) The human genome project. prospects and implications for clinical medicine. *JAMA* **266**, pp. 1966-975.

Harvey, W. (1628) *Exercitatioanatomicade motu cordis et sanguinis in animali,*. Fitzer, Frankfurt.

Kan, Y.W. and Dozy, A.M. (1978) Antenatal diagnosis of sickle cell anaemia by DNA analysis of amniotic fluid cells, *Lancet* **2**, pp. 910-912.

Landner E.S. (1996) The new genomics: global views of Biology, *Science* **274**, pp. 536-539.

Lips, C.J.M., Vasen, H.F.A., Lamers, C.B.H.W. (1985) Multiple endocrine neoplasia syndromes, in Hj. Müller, W. Weber (eds.), *Familial Cancer*, Karger Basel, pp. 103-111.

Ludman, M.D., Grabowski G.A., Goldberg J.D., Desnick R.J. (1986) Heterozygote detection and prenatal diagnosis for Tay-Sachs and Type I Gaucher Disease, in T.P. Carter, A.M. Willey (eds.), *Genetic Disease, Screening and Managemen*, Alan R. Liss, New York, pp. 19-48.

Martin J.B. (1993) Molecular genetics of neurological diseases, *Science* **262**, pp. 674-676.

McKusick, V.A. and Amberger, J.A. (1993) The morbid anatomy of the human genome: chromosomal location of mutations causing disease, *J. Med. Genet* **30**, pp. 1-12.

Müller Hj. (1990) Dominant inheritance in human cancer, *Anticancer Res.* **10**, pp. 505-512.

Müller, Hj. (1996) Genetic counselling of cancer patients and their relatives, in Hj. Müller, R.J. Scott, W. Weber (eds.), *Hereditary Cancer*, Karger, Basel, pp. 162-172.

Müller, Hj. (1993) Predictive genetic testing: possibilities, implications, limits, in H. Haker, R. Hearn, K. Steigleder (eds.), *Ethics of Human Genome Analysis*, Attempto Verlag Tübingen, pp. 136-146.

Thalhammer O. (1975) Frequency of inborn errors of metabolism, especially PKU, in some representative newborn screening centers around the world. A collaborative study, *Hum Gent* 30, pp. 273-286.

Vesal, A. (1543) *De humani corporis fabrica*, J. Oporinus, Basel.

10 Genomic instability – a story of repair, cancer and evolution with existential impact on the individual

NICOLAAS G.J. JASPERS
Molecular geneticist
Department of Cell Biology and Genetics, Erasmus University Rotterdam
PO Box 1738, NL-3000 DR Rotterdam

Further keywords
awareness, cultured cells, DNA damage, fate, functional domains, gene-targeting, inherited disease, materialism, mechanistic schemes, model systems, mutants, organismic level, predictive medicine, pre-embryo, presymptomatic diagnosis, radiation, reductionism, self-fulfilment, sequence conservation, soma-to-psyche shift, sunlight, toxicology, transgenic mice, ultraviolet, xeroderma pigmentosum

Abstract
The subject of inherited DNA repair deficiency in man is chosen as a typical example of how to investigate complex biological processes by modern molecular-genetic reductionistic approaches. The excision repair process is now on the verge of being 'elucidated' in terms of genes and proteins, resulting in a final 'mechanistic scheme'. While excitement about this accomplishment appears justified, the limitations of this knowledge become apparent as well. Ways to tackle the problem of 'biological relevance' are being sought, on the one hand, by extending and upgrading genetic technology to this level. This route entails the development of transgenic animals – its application will become widespread and its success in providing new insights far-reaching. Future expansion to the pre-embryo as a 'human system' is not excluded. Working with higher organisms is expected to increase our awareness of the 'organismic dimension'. Particularly with respect to man, genetic technology is now confronting us with the necessity to develop new morals, fitting to the rapid current changes in individual consciousness and responsibility. For instance, genetics have revolutionized predictive medicine which turns out to drastically dramatize the significance and experience of health and sickness, in social as well as individual life. Molecular genetics can be looked upon as just one exponent of a shift in attention from the somatic to the psychological level. If handled humanely, genetic technology can act as a powerful catalyst in this exciting and inevitable development.

I Introduction

Questions to be asked

In the previous presentation by Hansjakob Müller you have heard how the clinician is dealing with genetics and inheritance. Essentially, he talked about management: He described a flow of information, from laboratory to physician, and its consequences for patient care. He provided an overview of how many human disorders, with how many human disease conditions genetic management is currently feasible and applicable.
Indeed, on the final day of this meeting we have come to the topic of man: we do no longer discuss simple bacteria, plants or animals – now we discuss man and find out that we are facing very specific problems; the problems encountered in genetic technology and engineering of plants and animals are clearly distinct from those experienced with man. I hope to show you some examples of this.
What I would like to do today is view the subject of genetic technology from two different angles. On the one hand, I shall have to travel along Müller's path, but descend much more deeply into molecular analysis and pick a subject I think I know something about, i.e. the work I am currently involved in, back home in my laboratory in the Netherlands. You are allowed to immediately forget the actual results I will present to you today – instead, please read between the lines and take home with you its spirit, or the metaphors I use, or the fun of it, for instance. Looking from another angle, I would like to come back to the question that was raised yesterday by a lady in the audience. She asked "What the hell are we doing?" That is certainly a proper question to be asked and I will try to address it as well.

II Modern 'reductionistic' molecular genetics

Choice of subject

When I came into the lab in 1975, there was great interest in the biological effects of radiation. The dreadful events that ended the last world war, the deployment of vast amounts of nuclear weapons in the sixties and the serious attempts to exploit nuclear energy economically on a large scale, had made radiation research a very important issue. In the early post-war years, when most research had started with the simple organisms like E.coli and yeast, it was soon realized that not only radioactive material can cause damage. Also our beloved sun, the ultimate source of all our energy, joy and health, was found out to cause damage to bacteria, or to be more specific, to the DNA of bacteria. If DNA is isolated and exposed to the ultraviolet component of sunlight it will become damaged.
The typical approach of a geneticist to tackle any kind of problem, including that of the response to UV-light, is to look at the abnormal in stead of the normal. An abnormal situation tells us something about what is normal: so geneticists always start with a

search for mutants, crippled variants. Usually, most of such variants had been created in the lower organisms by the investigators themselves, by feeding the organisms all kinds of poisons. These mutants have given us the first ideas how the responses to UV come about. Subsequently researchers asked themselves the question: "Can such crippled variants also be found in man?" The answer is yes: it comes in the form of an extremely rare inherited disease – and I will just discuss this single disease today. It occurs at a frequency of less than 1:100,000 so there appear no implications for public health at all. Moreover, the disease cannot be treated and I do not expect it to become ever treatable even after extensive molecular research.

'Children of the dark'

The disease is called Xeroderma Pigmentosum (XP) – this name means 'dry skin with irregular pigmentation spots', a skin found exclusively on the body parts exposed to sunlight. The patients are really extremely sensitive to sunlight: when staying outside – even in the shade for only half an hour – they develop a serious sunburn reaction. So, without sun-protective measures, they have to remain indoors. The type of skin that is sometimes seen in older persons having experienced life-long intense sun exposure (farmers, ferrymen, roofers), such a severely keratinized, heavily pigmented 'old' skin is seen in XP patients already at a very young age, sometimes even in babies. Soon, many lesions emerge that will readily develop into cancer. The children have a 1000x increased skin cancer risk: usually, an XP patient has to have 3-7 suspicious lesions surgically removed from the skin each year – sooner or later there is one lesion too much. Other severe problems are seen in the eye, our light-dependent organ: photophobia (= fear of light) and blindness due to clouded lenses. Frequently the nervous system and brains are involved as well, causing serious and progressive disabling and mental deterioration.

DNA repair

Chemical studies had shown that two neighbouring pyrimidines in the DNA are specially 'vulnerable' to UV and can become linked by exposure: a so-called cyclobutane-pyrimidine-dimer is a prototype modification. Such molecular 'lesions' interfere with DNA function and are produced in massive amounts. In the case of human exposure: an ordinary day of sunbathing on the beach produces 10,000-100,000 of them in each cell of the skin, by far our largest organ. This damage has to be removed: it can be removed indeed, by a very clever process; from studies in bacteria is was already known how this grossly works.

Geneticists are used to draw DNA as a straight line to make it very simple (which it is not of course); the damage (schematized by a dot in the line of Fig. 1) is first recognized and then cut out from the DNA strand by specific enzymes, leaving a gap that can be filled-in again by polymerizing enzymes, using the intact homologous strand as a template. This filling-in process can be used to measure the course and extent of the process, now called

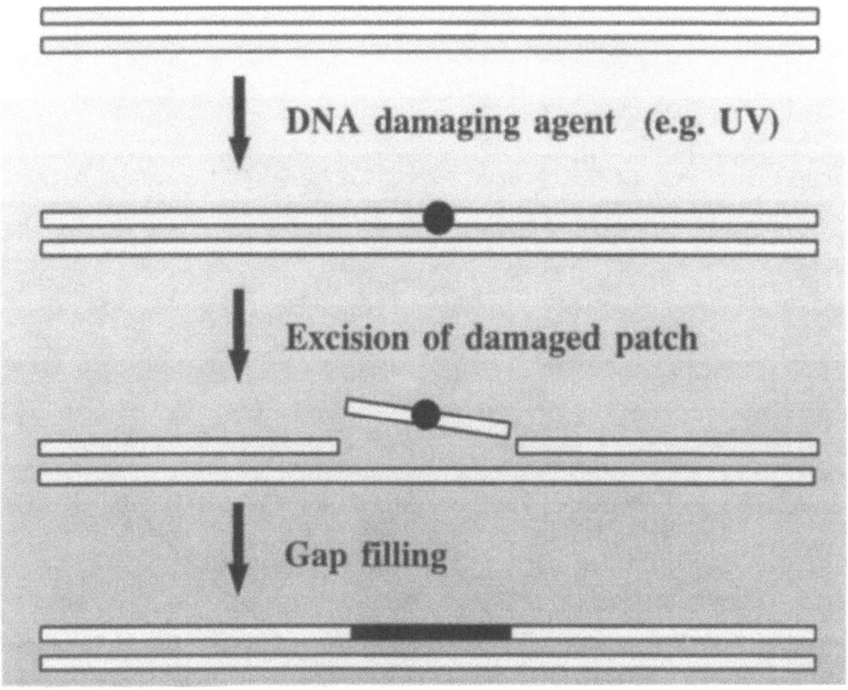

Figure 1. Basic principle of excision repair.

A damaging agent like UV-light causes a 'lesion' (•) in one of the two DNA strands. Excision repair works according to the cut-and-patch principle also used by plumbers and surgeons: the damaged piece is cut out and replaced by a new piece.

excision repair. If it is performed in the presence of radioactively marked DNA building blocks, the repaired DNA will become 'labelled' as well: repair activity will leave a radioactive trace and it's this trace that can be measured by all sorts of devices. If cells are repairing DNA in presence of radioactive DNA building blocks, they will become radioactive themselves.

From patients to cells *['down']*

The use of cells in stead of live persons in the study of DNA is evident. Frequent exposure of persons is unethical or at least impracticable, *a fortiori* in XP patients. However, the use of cells implicates a first but important 'reduction': we have restricted our investigations to the plantlike characteristics of man: we disregard typical animal properties like movement, emotions, pain and human 'things' like morality, freedom etc. In cultured human skin cells, DNA repair can be visualized as little black dots ('grains', *see*

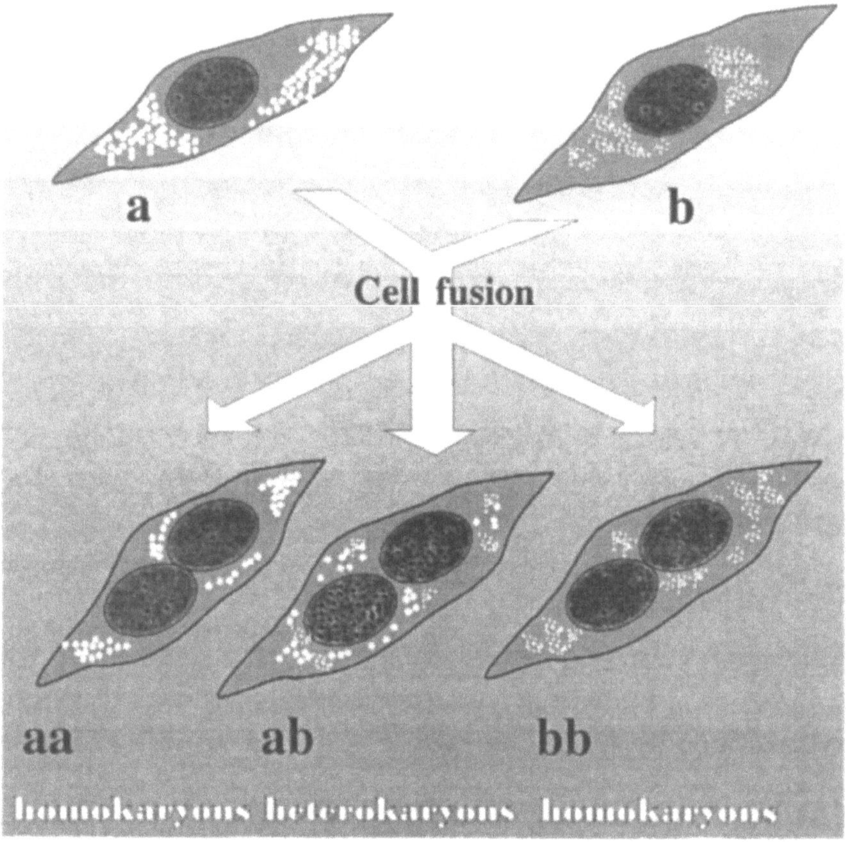

Figure 2. Genetic complementation analysis by somatic cell fusion.

At the top, two UV-irradiated fibroblasts in culture are shown from two different XP patients (a and b). The cytoplasms of the cells had been loaded with plastic beads, large ones in cells from patients a and smaller ones in those of b. In the nucleus, a few (2 or 3) black dots ('grains') are visible, representing repair activity (very low in a and b) visualized by a photographic technique. The DNA in the nucleus of a normal cell would have many grains, indicating efficient excision repair.

Cells from a and b are randomly mixed and fused, causing cells with more than one nucleus. Those with two nuclei, as shown on the bottom, come in three types: aa, ab and bb, which can be recognized by the presence of the beads in the cytoplasm. In this case the cells from a and b complement each other resulting in a high repair activity in the 'heterokaryon' of type ab.

Fig. 2) by photographic techniques which detect the radioactive tracer. In unirradiated normal cells and in UV-irradiated cells from XP-patients such grains will not be found: XP-patients are not able to repair DNA damage caused by UV-light. Studies in material from many patients soon revealed that this problem was not equally serious everywhere and this seemed to correlate grossly with the severity of the symptoms. Could it be, that in different patients DNA repair was affected in distinct ways? To answer this question in a genetic way, one would need a cross between patients; obviously an impractible approach: patients are just too rare ... Fortunately, crossing can be mimicked at the level of cultured cells by cell fusion: another obvious reason for working with cells. This type of study, called complementation analysis (*see Fig. 2*), revealed that in many cases, cells from two different patients could help ('complement') each other: in the fused cells normal DNA repair was restored. This implies that two such patients have a distinct DNA repair problem: in molecular terms: different repair enzymes are affected in the two. Extensive complementation experiments revealed that there are at least seven distinct XP patient groups in this respect: they have been named by letters, XP-A up to XP-G. To some extent, the patients are distinct in a grossly corresponding way; some groups have more skin cancers or more neurological problems than others.

From cells to genes *['down']*

So, there must be at least seven different enzymes involved in the repair process; in genetic jargon: "XP genetic heterogeneity reveals a high complexity of excision repair". In the XP patients, different genetic 'defects' must exist. Next step is to identify these seven different genes. There are various ways to do this; I will mention only one: corrective DNA transfer. One takes 'normal DNA', that is to say, DNA isolated from a normal (non-XP) person's cells and fragment it to gene-size. The pieces are then 'fed' to many millions of cultured repair-defective cells and a few of those (about 1%) will accidentally pick up such a piece to place it in their own chromosomes, where it can be 'expressed'. If the piece corresponds to the gene that is affected, the cells will become corrected. After irradiation of the culture with UV-light only corrected cells will survive: all other UV-sensitive cells (having taken up a 'wrong' piece or nothing at all) will die. In the very few cells fished out this way, the transferred DNA fragment is traced and subjected to further standard molecular genetic analysis: sequencing, chromosomal localization, expression etc.

From genes to proteins – from man to yeast *['down']*

By translating the nucleotide sequence (just on paper!) into the polypeptide sequence of the encoded protein one can obtain a first clue to the actual function of the protein. Note that we have never 'seen' this protein or even studied it! By analysis of this so-called *putative* protein, some specific features ('domains') can be recognized that were seen earlier in the genes encoding other better known enzymes. *Fig. 3* shows the example of

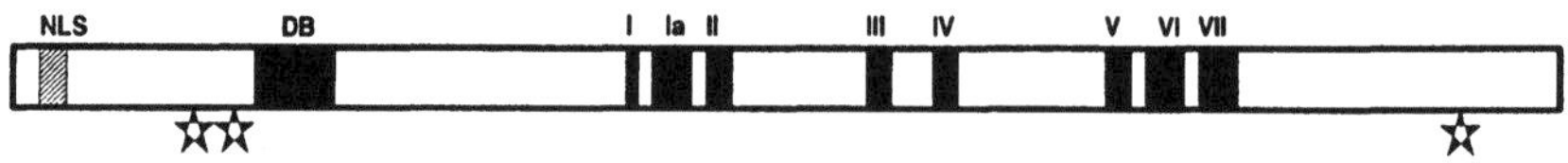

Figure 3. Repair protein XPB.

(A) Linear representation. Seven 'domains' of the sequence (dark grey) are highly conserved in enzymes with 'helicase' activity. NLS (hatched, 'nuclear localization signal') is another domain that assures routing of the protein to the nucleus. The DB area (black) confers 'putative' DNA binding capacity. The stars indicate a few sites where mutations occur in patients. In the 3-D (folded) structure of such enzymes (very recently resolved by crystallographic analysis), the 'helicase' domains are very near to each other, on the bottom of a furrow where the DNA molecule fits in and slides through, while being 'unzipped' at the expense of energy.

the enzyme encoded by the XP-B gene: here seven domains were identified which are characteristic for 'helicases', these are enzymes able to unwind the two complementary DNA strands. How such hunches can be verified, will become clear in a few moments.

The conservation principle *['down']*

What the sequence revealed further was 'homology': genes with very similar coding sequences exist in lower organisms and they encode very similar repair enzymes. The jargon says: "Excision repair genes are evolutionary conserved". This implies that even in simple fungi like the baker's yeast Saccharomyces Cerevisiae or a fully unrelated yeast called Schizosaccharomyces Pombe, excision repair works grossly in the same manner. That DNA repair has been 'invented' by nature a long time ago, was something to be expected: the sun has been around already for quite some time. But that the repair processes have remained relatively unchanged was not the first thing to be expected. What it implies is that we can use molecular genetic knowledge obtained on simple organisms to understand basic processes in man: these basic processes (remember) are restricted to the single-cell properties, the plantlike properties of man.

While the human genome project is advancing, identification of genes and proteins becomes easier every day: from now on many of the new 'disease genes' are picked up by computer! The conservation principle has taught geneticists the importance of not restricting the Human Genome Project to man; simpler prototype organisms like yeasts, flies (Drosophila), worms (C.elegans) or fish are included in order to help identification of relevant coding areas in the human genome. The latest DNA repair gene that was cloned may serve as an example. On the basis of peptide sequence similarity and functional resemblance, two yeasts S.cerevisiae and S.Pombe were expected to carry the 'candidate genes' for the homologs of the human XP-F gene. The most homologous (i.e. best conserved) regions in the two corresponding yeast genes were used to search for the 'human homolog'. The computer search was performed in part by a commercial company, which has made a vast library of human sequences and sells parts of it to anyone interested. This company provided the 'front part' of the gene. By our own laboratory work on a cloned library we identified the rear part of the same gene. These two partly overlapping gene fragments were assembled to one sequence with relatively little difficulty. The commercial bit however may sometimes be less easy: imagine how one has to divide up the profits and financial rights connected to one single gene if two owners claim overlapping halves...

From proteins to functions *['down']*

Now the involvement of this human gene in XP patients from group F had to be verified. DNA from a patient belonging to this group was sequenced and it turned out we were right: both alleles, the one from the patient's father and that from his mother, both contained a serious mutation. To be really sure, the 'normal gene' was injected into the nucleus of this patient's cultured cells, using a glass microneedle. The repair capability of the injected cells stood out positive amidst the surrounding uninjected cells. Knowing now that we had identified the XPF gene, we had to find out what the encoded protein can do – in other words we were left with the enzymatic function of the protein. How this was done in case of XPF may well serve as an example for the many other repair enzymes. From yeast studies the suggestion arose that this last excision repair enzyme probably had to do something with a central step in the repair process: making the

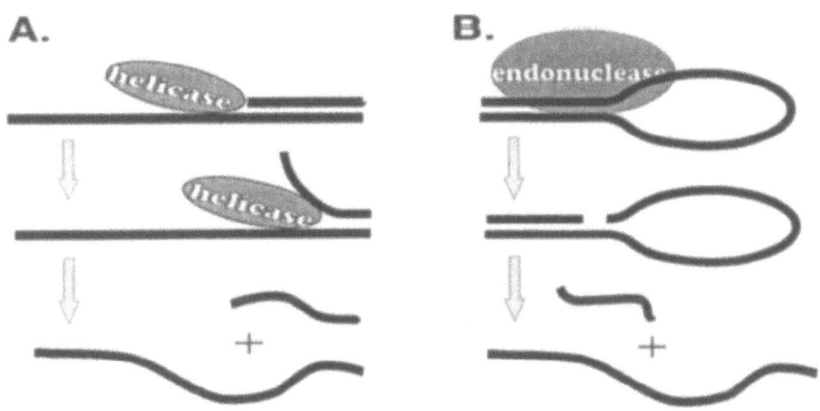

Figure 4. Artificial DNA molecules used for DNA repair studies.

(A) To measure 'helicase' activity. The short piece (carrying a radioactive trace molecule) is displaced from the larger one by, for instance, the XPB enzyme. (B) A stem-loop type of structure that can be incised by the XPF enzyme.

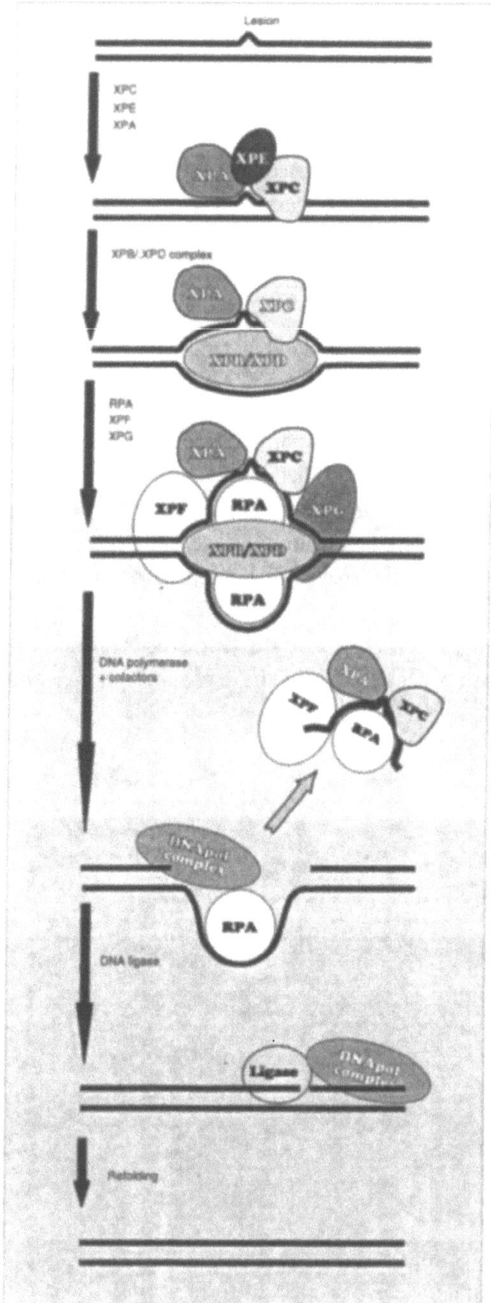

Figure 5. Hypothetical mechanistic scheme of the excision repair process.

The scheme is simplified, indicating the main proteins only. Five phases can be recognized:

[1] **Recognition.** XPC and XPA proteins bind to the damage, helped by XPE.

[2] **Unwinding.** XPB and XPD can in turn bind to XPA and then start to separate ('unwind') the two DNA strands, each in opposite directions. Another enzyme called RPA is also attracted to the site by its ability to bind to XPA as well as single-stranded DNA and starts to cover the opened molecule over a distance of 30 nucleotides.

[3] **Strand incision.** The damaged strand is cut on both sides of the lesion. The XPF enzyme is put in position because it can 'recognize' XPA, RPA and the strand transition and makes a cut on the left sid. The other cutting enzyme XPG is directed to the right side by recognition of the unwinding enzymes.

[4] **Resynthesis.** Starting on left side of the lesion, a new DNA strand is made by a number of 'standard' polymerizing enzymes which are also used for DNA replication (DNA polymerase epsilon plus some 'helper' enzymes). In the process, the damaged patch is believed to be displaced and subsequently digested by unknown enzymes.

[5] **Rejoining.** Finally, the open end of the new DNA strand is fixed to the existing end on the right side of the lesion. The 'responsible' enzyme is DNA ligase.

incision in the DNA strand next to the DNA damage. This idea we have tried to verify as artificially as possible. A computer-guided automat was used to synthesize a short piece of DNA with a 'stem-loop' structure (*see Fig. 4*) that looks very much like an intermediate in the still hypothetical excision repair reaction: it represents a partly opened DNA strand. This artificial substrate is added to an XPF enzyme that we have isolated from many billions of cultured normal cells (I'm afraid I'll have to spare you the biochemical and genetic tricks required to accomplish this hard job). The protein turned out to make an incision in the substrate indeed: and precisely at the site where double-stranded 'stem' switches to the single-stranded 'loop' (Fig. 4). In the jargon this sounds like: "the XPF excision repair protein is a structure-specific endonuclease", a message that we were quite excited about, and so were the editors and referees of the scientific journal "Cell", only very recently.

From functions to schemes *['down']*

So now we believe that we have elucidated basically *how it works* – we 'just' have to put the pieces together in an overall scheme, a 'model' for the 'mechanism' of excision repair. Fig. 5 shows how the main steps in the process could follow each other. It is really fascinating to see how all the genes affected in the variety of XP patients fit in. Even more exciting, to me as a scientist at least, is the recent accomplishment in British and American laboratories, that all the 20-30 different human repair enzymes, mixed together in a test-tube after separate preparation and isolation, actually prove capable of removing DNA damage from UV-irradiated DNA added to it. Only a few years of hard work will be required to actually prove (disprove?, adjust?) the hypothetical scheme of the figure and to deal with the details of the process, for example, which proteins interact and how the damaged DNA is refolded. Subsequently, the (more complicated) task is to find out how the repair is regulated and fine-tuned, since it is known already that some regions of the genome (notably those which are under active transcription) are repaired by cells more efficiently than others.

From schemes to organisms *['up'?]*

Now *what* have found? Well, we now *how it works*, one would say, fair enough, because that's the reason why it was done. We believe to have *solved the problem* on the molecular level, but we are wrong. We should not forget that we have just cells here, mutant cells, very different form organisms or patients. In one way or another we have to get back to this level. The molecular level of DNA is about the lowest you can get to, in order to have an understanding of life – the most basic if you wish: 'you cannot land deeper in the gutter than to work with genes' criticists might say, and in a way, this is not a joke at all but really true: abstract gene sequences may represent the furthest point of reductionist biology indeed. A way-out to this well-recognized problem in biological science is offered by genetically modified mammals, in particular mice. This has been

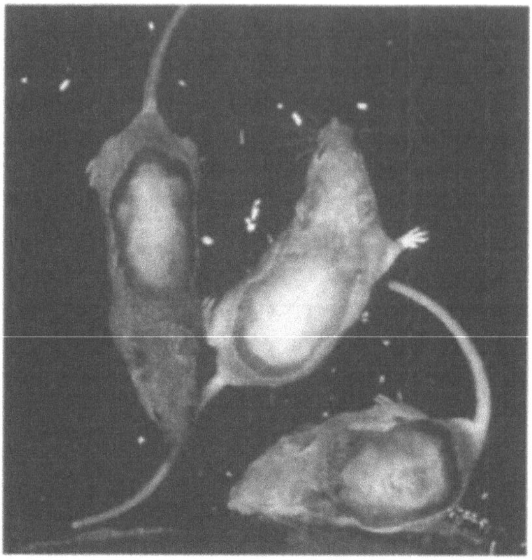

Figure 6. Mice with a modified excision repair gene.

These three littermates are the outcome of a 'gene targeting' experiment with an excision repair enzyme. In one of the three mice both of the repair genes (that from the father and the mother) were left unmodified, one of them is a *heterozygote*: only one of its two genes is affected and in the third mouse both genes were made defective (a *homozygote*). The fur on the back of the mice was partly shaven to expose them to a low dose of UV-light from a sun lamp. In one of the three (the one with the 'double knockout') there is a furious skin reaction, whereas the other two animals can easily cope with the given exposure.

accomplished (and surely not only by colleagues in my lab) in the study of DNA repair too. On the picture (Fig. 6) you see three mice from the same litter. One is healthy (in terms of genes), the other carries a mutated repair gene in one of the two alleles (it is called 'a heterozygote') and the third, has two mutated alleles of the repair gene (it's a homozygote). When these mice are exposed to a small dose of UV-light, a dose that doesn't harm the healthy and heterozygote sibs at all, the homozygote shows a furious skin reaction. And if this UV-treatment with a very much lower dose is repeated daily, all of the sensitive mice will develop skin cancer, and very soon. However, as long as they are kept away from the UV-lamp, they all appear to function normally (what is normal in the artificial conditions of the laboratory?) – that is to say, their behaviour appears not to differ from any unmodified animal. In fact, these mice represent 'a good animal model for XP'. They will probably prove to be highly valuable in elucidating many of the separate steps of skin carcinogenesis, because this can now be studied at what we call 'physiological doses', the UV exposures received by not too excessive sunbathing. In addition, these mice may help us to identify and trace the action of DNA damaging agents other than UV (cigarette smoke for instance): they will be of toxicological relevance as well. On the other hand, the mouse studies have also presented us challenging puzzles. The 'knocking out' of another repair enzyme resulted again in highly UV-sensitive mice, but now the homozygote animals were much smaller than their littermates and seriously ill. Within a few weeks, they died with failing livers and kidneys and many other problems. And there are now more examples where repair-deficient 'knockouts' do not (fully) match their human counterparts.

So one time, there is 'a model' for the human disease and the other time there is not – at least not in the way we expected from the simple single cells in culture (yeasts, mammals). The implication is that this route back to the organism, as expected, will provide us an entirely new level of knowledge. No wonder, that in present science, this route is extremely popular: hundreds of different mouse models have already been generated, in many cases with fully unexpected outcomes. The trend in the near future will be that (1) genetic modifications ('gene targeting') will become easier to do, more sophisticated and more subtle than the simple and straightforward 'knock out mutations' that fully inactivate the function of the genes. Furthermore (2), when increasing numbers of genes will become identified by the human genome project (in total there are some 60,000 to discover in the coming decade) a large part of them will have to 'targeted' in the mouse in order to have some idea of the role(s) they play in physiology. In some cases the animal may seriously suffer, in many cases the mutations will either be lethal (no mice born at all) or fully 'unharmful', i.e. without any effect. This new way of handling experimental animals will certainly increase awareness of animal suffering and will create novel attitudes towards livestock – there will be a lot to do for the animal protectionist movements; for the sake of the animals, I do hope that they will have the wisdom to follow a strategy of dialogue, one of listening to each other, as we have been doing here latest days.

Back to the organism: the human case *['up'?]*

The obvious drawback of these models is the species difference: mice and men are really not the same, as everyone knows. As a consequence, the future will hear stronger calls for similar strategies to be applied to man. This cannot immediately be done with people like you and me, of course; to this end, a *reduced* human being is required. A likely candidate is already presenting in the form of the pre-embryo. The pre-embryo (many embryologists tell me) probably does not represent something physiological, something 'real'; it rather appears an creation-of-thought, a concept, based upon and in line with the current widely accepted moral on 'gestational management' and abortion. The pre-embryo is not considered as a *human being*, by definition it is a *human system*, a disposable one – just like an organ or cultured cells. As such it will probably be the first human 'organismic' level for experimental molecular biology to venture on in the future.

Currently, there is no need for it at all – so much fundamental knowledge (relevant for man as well) is still to be discovered in mice. A general, international moratory on such experiments is in effect. However, I doubt whether this moratory can sustain for another decade. A number of investigators now already openly advocate pre-embryo studies.

III Perspectives

The human case revisited *['up'?]*

Coming to the end of this exposé, it is about time to address the question, what it is that we are really doing with genetic technology. The ways to return to the organism mentioned above represents the 'materialist' route, the one that becomes apparent when physical and chemical aspects are the only themes to consider. Especially in the human case, we are forced to extend our scope to cultural and social levels. Unable to discuss everything here, I must restrict to a few simple examples, where practice turns out to be the best teacher – as usual. I remind you of the patients with serious inherited disorders. What genetic technology has offered primarily here, is an immense progress of diagnostic power. In contrast, progress in terms of somatic gene therapy is expected to be extremely moderate and, in the long run, effective only in a few diseases. However, it is out of question, that DNA analysis will have a steadily increasing significance for predictive medicine. It enables to make diagnosis on a very young age. For the parents and relatives of the patients with incurable and lethal disorders, the DNA tests have a strong impact. After an initial reaction of despair (why us? what to do?) there comes a feeling of relief. Suddenly the strange pains and complaints, that none could ever understand, (sometimes parents are accused of mistreatment!) finally have a name and a 'measurable cause'. And, more important, the future perspectives ('the prognosis') become clear. The predictive power of these tests is great. The parent now will become aware that there will be no real cure, that their child will die, very often slowly and painfully and with neurological deterioration. But this awareness in most cases is metamorphosed into something new. Parents (and patients) find others in the country, in the world, who share this fate. They come together and correspond in many ways to help each other or to put pressure on medical and social institutions to do more for their children. This second phase, which is seen every time again, is of particular interest: *knowing about the future, is knowing what to do.* There is a new way of concern, a way to love the children even more – it's not them where the problem is, it's only in their genes.

Being judge of your own future

But why do we want to know all this? I believe this is a typical trait of our time. We want to self-fulfill, we want to be the judge of our own future. We are highly interested in our future – more than we ever have been before. This is just one example, that may tell is what new genetic technology is really doing. Predictive medicine will therefore receive growing attention. Its existential impact is enormous in case of presymptomatic diagnostics. Women carrying a defective BRCA1 gene, which predisposes them to suffer from an aggressive form of breast or endometrial cancer at a young age (30-35 or so), are now faced with deep-cutting existential problems after a genetic presymptomatic diagnosis: how to make decisions for careers, decisions for preventive amputation of breasts and ovarium and when if at all? Do I try to have children first, how do I live with the continuous threat, etc etc. This is what predictive medicine does: the fate of the individual carrying a disease predisposing gene has changed in a profound way. Before it, the fate was experienced on the somatic level – eventually, the woman has to suffer from breast cancer, invasive and painful therapy regimens and (too often still) death. Now the fate is experienced much earlier and the suffering has shifted to the psychological level. Disease experience has moved away from the affected organ in the direction of the mind. There is no way to escape the existential decisions, as soon as the technology becomes available: to the least, one is forced to make a choice between the wish 'to know' or 'not-to-know'.

Being 'a characteristic of our age', these developments in predictive genetic diagnostics are not at all standing on their own. Numerous examples on other fields can be found in addition to other genetic disorders (Huntington's disease). As an example of acquired (non congenital) fatal disease there is the example of AIDS: I remind you that the period being seropositive without clear symptoms lasts usually much longer than the actual disease itself. However, already in this presymptomatic phase the impact is serious and affects individual life and personal contacts, extending up to the macro-social level, in terms of discrimination, insurance policy etc. New types of needs and helpers ('buddies') as now emerging with AIDS, will probably become important for genetic disease too. Another striking example recently came to my knowledge from the newspapers, highlighting that the Chernobyl disaster has happened ten years ago. It turns out that the psychological and socio-cultural distress caused by the nuclear reactor accident vastly exceeds the suffering that could ever be created by actual radiation disease itself – which is certainly evident as well.

My personal believe is, that genetic technology has not developed simply as a logical sequence of events in a linear scientific progress. We, our society, have acquired means for genetic technology because they fulfill our fundamental needs. And in our time (at least in the rich countries), these needs are on the level of personal fulfilment (Table 1 provides a summary). GeneTech is expected to gain importance simply by this fact. At the same time, being essentially an informational trend, it fits very well in even wider and faster revolutionary developments of digital informational technology. GeneTech can function as a help to increase our awareness of present and future and of the real needs in personal and social life, something we almost desperately look for.

Table 1

FROM ORGANISM TO MOLECULES - AND BACK : TWO ROUTES

Route Description	Signature	Trends	Outcome
Physico-chemical example: gene targeting	'materialistic' the organism as a highly ordered and complex array of molecules	PAST: mutant cells, lower organisms FUTURE: transgenic crops, animals pre-embryo?	[Factual] [a] a. insight, know-how b. technology, management c. leasure, prosperity, power
Psycho-social example: predictive medicine • in wider perspective: - cancer - neurodegenerative disorder - AIDS - Chernobyl suffering	'spiritual' organisms with 'intrinsic value' as a species or as a person	*"from soma to psyche"* PAST: somatic experiences *(pain, disability, death)* • therapy = institutional medicine FUTURE: psychological experiences *(distress, changing perspectives)* *(mutilation, stigmatisation)* • therapy = individual loving care	[Positive] a. awareness, consciousness b. care, solidarity c. self-fulfilment, freedom [Negative] a. anxiety, denial b. isolationism, bureaucracy c. predictability, repression

Note: (a) negative and positive aspects of technology in general are considered as widely recognized here and therefore not mentioned.

11 Human biography and its genetic instrument

MICHAELA GLÖCKLER
Physician
Medical Section at the Goetheanum
PO Box 134, CH-4143 Dornach

Abstract
The aim of this paper is to demonstrate the significance of some of the results of Rudolf Steiner's research for modern developments in genetics. The key concepts are: the etheric organism as the bearer of the laws and processes of inheritance; the metamorphosis of growth forces into those of thinking; the human 'I' (self) as controller and modifier of genetic material and self-organisation (molecular-Darwinism) seen anew in the light of anthroposophy.

Introduction
Human biography goes through characteristic stages in body, soul and spirit. At the physical level, active growth and development up to age 20 or 25 is followed by physiological functions continuing at the level reached until about 40 or 45, and then progressive involution and the physiological deterioration of old age to the end of life. Apart from this there are characteristic pathological processes, with acute febrile infectious diseases at their highest level in childhood, chronic diseases in old age, and psychosomatic conditions in mid-life. Compared to this, the human biography in soul and spirit shows remarkable differentiation. Childhood and youth show considerable differences depending on where a person grows up, what kind of schooling he or she has and where his or her interests lie. Choice of occupation and working life, the circumstances of private life – with or without family – all provide highly differentiated opportunities for learning and experience to further individual development.

Inheritance, environment and individuality
One of the most interesting questions to have arisen – once the theories of evolution and heredity gained general acceptance in the 19th century – is how far human development and hence also biography are predetermined by genetic or environmental factors and by 'personality' – a rather vague term – or the human "I" or ego. In their

book 'Separate lives. Why siblings are so different' (Dunn and Plomin, 1990), developmental psychologist Judy Dunn and behavioural geneticist Robert Plomin discuss that question in the light of their extensive researches. They analysed research findings made in developmental psychology and behavioural genetics in recent decades and a large number of studies and surveys to come closer to an answer. A remarkable discovery they made is that similarity in terms of size, weight and disposition to diseases, for instance, is rarely greater than 50% and generally well below this. Thus differences between siblings concerning distance between eyes, length and width of nose and length of ears is at around 30%. About 80% of all siblings have distinctly different eye colours; 90% differ in the colour and structure of their hair. The disposition for diseases such as gastric ulcer, hypertension, breast cancer, diabetes, childhood eczema and for asthma and hayfever is at less than 20%. These findings are not in accord with modern concepts of heredity, for it has been shown that surprisingly few behaviour patterns can be ascribed to a single gene. It is generally the case that several genes are responsible for a particular characteristic, and it may be a case of hundreds of genes each making a minor contribution to variability between individuals. The resulting genetic effects are called 'additive effects'. Non-additive effects arise when the influence of genes changes due to their particular position or even their mere presence, creating new characteristics that have not existed before.

Dunn and Plomin made use of the term 'epistasis', originally introduced by biologist William Bateson in 1907 for such non-additive gene effects leading to the appearance of new, unexpected characteristics in individuals. They use the term for genetic effects which, being part of the interaction of genes, result in spontaneous, unpredictable interrelationships in the highly complex dynamic network of genes. In their view it is also due to these effects of a higher order that even first degree relatives show less then half the similarities of identical twins. According to these theories, therefore, similarities between offspring is due to additive summation of numerous individual genetic effects, whereas differences between offspring are the result of epistasis, i.e. those effects of a higher order, causing non-additive and therefore unpredictable new characteristics.

Environmental factors, often difficult to define but needing to be differentiated, also come into this. They, too, can mutually enhance, weaken or balance one another. Depending on their degree and quality they can change the genetic material of plants, animals and humans, and this means a further vast range of additions, interactions and potential enhancement, weakening or balancing out that cannot be individually predicted or indeed assessed. In a seemingly 'chance' way, additive and epistatic effects in the genome combine with those due to environmental factors, proving 'lucky' or 'unlucky' in the spectrum of the individual's gifts and limitations.

The question is, however, whether the terms 'chance', 'luck' or 'bad luck' adequately define the principle according to which a combination of genes is selected when ovum and sperm fuse, then to be influenced by environmental factors with their innumerable additive and epistatic effects.

The question is taken to an even deeper level because Dunn and Plomin's work has brought together masses of material that add individual relationships as a third key concept to those of heredity and environment. Analysis and systematic evaluation of numerous studies, including the study and comparison of many writers' biographies and childhood memories, have shown that a child's behaviour is not solely determined by heredity and environment, but that it also depends to a major extent on the child itself how it deals with the environment. The surprising result of these investigations has been that children themselves are remarkably selective in the attention and capacity for relationship they have for those environmental factors. The key aspect of potential environmental effects was always the existence and nature of individual relationships that really allowed the environment to have an influence, aspects that gain in power or lose significance for individual development. If heredity and environment were the only influential factors, siblings would show much more similar behaviour than they actually do. The real situation is that siblings also differ because their relationships to one another and their parents are vastly different, shaping them individually. Relations with the same parents can be so radically different that an outsider hearing of them can only conclude these must be different people. The same mother may be seen as warm-hearted and protective by one child, and impatient and daunting by another (Dunn and Plomin, 1990). Irrespective of whether parents think they are essentially treating their children the same, or of their idea of preferential treatment or rejection, children are very accurate in saying how they do or did experience parental attitudes, and their perceptions will often differ greatly from those both of siblings and the parents themselves – they are highly individual.

Development and cosmic order

The hypothesis of chance as prime mover in development grows even more doubtful when we consider the numerous correspondences and connections human biography shows in relation to the evolution of the realms of nature and the situation of the whole cosmos. Human metabolism follows circadian rhythms which in turn depend on the sun's orbit around the earth. Pregnancy is still counted in lunar months, embryonic development proceeding in characteristic weekly and 4-weekly stages, not to speak of the direct effect of the sun on the vitamin D precursor in human skin, for instance, and on photosynthesis in plants. In terms of space and of rhythmic time sequences, earth and cosmos are part of a finely attuned system of relationships; nothing drops out of this as being 'meaningless'.

This goes so far that even orders of magnitude show exact correspondence. I once worked out where we would get to if we took the number steps in powers of ten into the universe that we need to take to come to the smallest building stones of matter, which are in the range of 10^{-16} – 10^{-18} meters. 10^{-16} – 10^{-18} meters reflect the distance to the nearest fixed stars, e.g. Alpha Centauri.

To speak of chance when one knows of all these interactions and correspondences resulting from genetics, the influence of environmental factors and finally also the whole of nature and the cosmos around us, is to my mind to speak of a hypothesis that rests on extremely shaky foundations. I would therefore like to oppose this with another hypothesis that comes from Rudolf Steiner's science of the spirit. It says that human thinking is capable of more than discovering all the above relationships and correspondences and formulating laws of nature for all visible phenomena. Our thinking is the evolutive power that is active in the vast variety of species in the plant, animal and human worlds. Thinking activity itself is a power that has emancipated from natural processes and has independent, purely spiritual evolutive powers. This requires elucidation, and a brief outline is given below (Steiner and Wegman, 1925).

Comparison of human and animal development

In the first place it is remarkable that compared to the different animal species, human beings are less specialized. Comparative study of human and mammalian embryonic development shows the morphology to be practically the same, phenomenologically speaking, in the early stages. Even a bird's wing looks like a human arm in the early stages of development, with five finger rays. This form is abandoned as development progresses, with the fingers growing stunted, fusing in some areas, and the central ray extending, finally producing the highly specialized skeleton of a flying creature. It can come as quite a shock to compare the head of a chimpanzee immediately before and after birth with that of a human baby. In form they are extraordinarily similar – small nose, bulging forehead, beautifully shaped back of the head. As the young chimpanzee develops, this vertical arrangement of the face is given up, projecting brows develop, the forehead slopes backwards and the back of the head loses its roundness. A characteristic ape's head results, with the facial part of the skull projecting (Poppelbaum, 1931; Kipp, 1980).

Human beings retain their non-specialized body form for life. One cannot tell by looking at a human hand if it is about to caress, strike, raise a dagger or play the piano. No specific use has been laid down for it and direction has to be given to its actions by specific thoughts about what is to be done next. This phenomenon shows a radical difference between humans and animals, a difference not sufficiently regarded by biologists today. Human beings have to make up for their lack of bodily specialization, and the lack of instinct-guided behaviour due to this, by using their intelligence. It means that animal intelligence is guided by body and instinct, so that animal behaviour is perfect and always optimally adapted to the ecological system. Humans on the other hand are the only creatures in the natural world to have an abstract intelligence that is not guided by body or instinct but has to be guided by the mind. On the basis of his spiritual researches, Rudolf Steiner explained this by saying that the same evolutive power that produced the different animal species and reflects in wisdom-guided species-specific animal behaviours, can emancipate from the body in humans, so that the body does not reach full specialization and human beings have that characteristic uncertainty relating to their instincts. This power is instead available as the thinking system that enables humans to make up the instinctual 'deficit' in full conscious awareness. The hypothesis shows why there is no law in nature that cannot be thought by the human mind and why human intelligence corresponds to all the laws according to which the human body is built and according to which the configuration of the human body relates to all other natural entities and the whole context of earth and universe. The hypothesis also shows why creation of the natural world ended with the animals and a transition occurs with human beings, natural development becoming development in mind and spirit. Humans grow out of what is given in the world we perceive with the senses and into developmental contexts that can no longer be perceived and grasped by the senses. Humans are able in their thoughts to set goals and ideals for their biography and their life, working to bring them to realization in their lives, though they may never bring them wholly to fruition. They have greater – spiritual – developmental potential than can be realized in a life.

Human biography and spiritual activity

I would now like to take up the theme of this essay – human biography and its genetic instrument. The essential armamentarium for human biography is human thinking. This gives orientation as to how to proceed, setting goals for development. Psychoneuroimmunological research and psychotherapy for serious immunological diseases such as cancer and AIDS have shown quite clearly that the functional capacity of the immune system depends to a high degree on a motivated, positive approach to life, some degree of idealism, and the ability to set goals (Adler et al., 1991). Thus there is direct interaction between human thinking and the biological processes in the body. This relates not only to immunological processes but also to the whole of human vitality, and especially also the reproductive system. Our genetic material is not only influenced by environmental factors today, but also by the way people think, feel and shape the life of the conscious mind.

A great deal of research will clearly be needed to verify the details of the working hypothesis so clearly stated by Rudolf Steiner. Let me show it in a sketch here in conclusion.

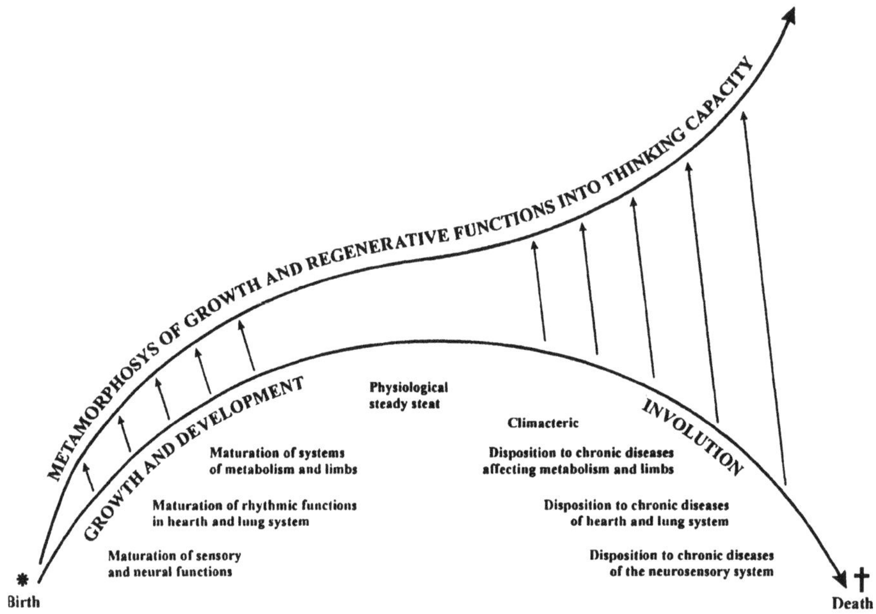

Figure 1: Schematic representation of human biological and spiritual development during biography.

According to this hypothesis, genetic material is selected not on the basis of good or bad luck or random chance, but by the human I, a reality in the spirit even before birth. With the aid of its thought organism (called the 'ether body' by Rudolf Steiner, 1904) it has selected and combined the paternal and maternal genetic material in the way most appropriate to the individual's development. The genetic piano keys of our ancestors are played by the essential individual nature of the human being himself, and this also takes in the environmental influences it needs by developing specific interest and abilities to relate. Using its powers of thought, individual human nature sets its own developmental goals. At the same time it will depend on how it thinks about nature and the earth how these will develop in the centuries ahead. For in the human being evolutive potential comes free of the germ track of species, being available to man as thinking activity, independent creative potential to develop a concept of his own further development.

References and notes

Adler, R., Felten, D.L. and Cohen, N. (1991) *Psychoneuroimmunology*, Academic Press, San Diego, NY.

Dunn, J. and R. Plomin (1990). *Separate lives. Why siblings are so different.* New York.

Kipp F. (1980), *Die Entwicklung des Menschen im Hinblick auf seine lange Jugendzeit*, Verlag Freies Geistesleben, Stuttgart.

Poppelbaum H. (1931), *Man and Animal – The Essential Difference*, Anthroposophical Press, London.

Poppelbaum H. in: Bockemühl e.a., eds. (1985) *Toward a Phenomenology of the etheric world*, Anthroposophic Press, New York.

Steiner R. (1904) *Theosophy. The essential nature of man.* Anthroposophic Press, New York.

Steiner, R. and Wegman, I. (1925) *Grundlegendes für eine Erweiterung der Heilkunst*, Verlag der Rudolf Steiner Nachlassverwaltung, Dornach. English translation: *Extending Practical Medicine* (or *Fundamentals of Therapy*, in earlier translations), 1996, Rudolf Steiner Press, London.

Acknowledgements

Anna Meuss' English translation, Johannes Kühl's critical editing of the manuscript and Georg Iliev's help with figure 1 are greatly acknowledged.

12 Practising a power free dialogue in the plenary sessions about modern biotechnology

DISCUSSIONLEADER: HENK VERHOOG
Theoretical biologist
Institute for Evolutionary and Ecological Sciences, Leiden University
PO Box 9516, NL-2300 RA Leiden

Reporters: Henk Verhoog and Arno van 't Hoog (NL- Leiden)

Introduction: Ethical individualism and pluralism

It has been the explicit aim of the organizers of this *If gene* conference to create possibilities for a real dialogue between people who are concerned about the 'future of DNA', the future of genetic engineering of living organisms. The processes of coming to a (moral) judgment are considered to be more important than the judgment itself. Ethical judgement can take place at different levels. Most important is the personal judgement of individuals in a particular social situation (ethical individualism in real life decisions). Important aspects in this connection are the person's biography and the tension between individual moral responsibility and social constraints (especially economic and political constraints). In the philosophy of *Ifgene* it is of the utmost importance to show a sincere interest in the life history of each individual, keeping back any moral judgements about the person in question. For the realisation of a power free dialogue it is important to strive after value-clarification: what is the person aiming at and why, what are the constraints under which she/he is working, why are certain choices made or not made? etc. An attitude of respect for the individual human being is required.

Two other levels of ethical judgement are the cultural and the political. *Ifgene* is mainly active at the cultural level of social life, not at the political one. At the cultural level, freedom and pluralism are important requirements; persons taking part in processes of ethical judgment should feel free to express their feelings and to give their personal opinion, free from any ideological, political or economic constraints. It is important to rise above the level of personal, economic or political interests and create a climate in which a power free dialogue can take place. What counts are the arguments used, not the person (or any authority) who argues. Arguments can be one-sided, biased, etc. and can be criticized because of this. Criticism is welcomed at the cultural level. Fundamental concepts such as truth and justice are necessary as regulative ideals. At the cultural level the equivalent of ethical individualism is pluralism. Personally I may be convinced that I know what is true or good with respect to a particular problem, but in a dialogue with others I must respect the freedom of the other participants to have their views. No dialogue is possible when participants claim that there is only one truth and that they possess it. This is not pluralism, but fundamentalism (absolutism). The other extreme is relativism, where the ideal of truth is given up, and the views of people are seen as relative to their social position. In extreme relativism dialogue is useless because people are believed to be unable to escape from the situation in which they are imprisoned. The cultural ideal of pluralism requires a fundamental openness to the ideas of others, willingness to listen to others and to revise one's own ideas if this is believed to be necessary in the search for truth or justice. Every idea can be questioned; dogmas are not accepted at this level. Pluralism is the cultural manifestation of ethical individualism; it is implied by the respect for the human being, for what it means to be human.

Ifgene's choice to be active at the cultural level, to stimulate the process of ethical judgment about aspects of genetic engineering, implies that *Ifgene* as an organisation has no political goals, such as a ban on xenotransplantation or on the patenting of living organisms. Participants in *Ifgene* are free, of course, to be active in the political field, to get their views heard in democratic decision-making processes, or to take part in coalitions. It is an important aspect of processes of ethical judgment that, although they are fully dependent on individual persons, the ethical intuitions arising in this process increasingly come from a level which goes beyond the individual. This may be compared with what in philosophical ethics is called the 'universalizability' of normative statements.
When you are convinced of an injustice in society, it is very natural that you speak up against this injustice. The reason behind this is that in spite of, or perhaps because of, our individuality we are human and we want our society to be 'humane'.

These ideas were constantly in the background during the plenary sessions. The issues discussed in the first two sessions were grouped together under the topic 'the role of science in our life-world'. The third session was about the risks of genetic engineering and whether there should be a moratorium; the fourth session dealt with the origin of our values and the final session focused on eugenics. Participants other than the main speakers are not mentioned by name.

Science and the life-world

The problem discussed during the first two sessions was most explicitly formulated in the lecture by Jaap van der Wal. He distinguished between the world of our immediate experiences, perceptions, feelings and values and the world of science, the world of objectified facts unrelated to our feelings and beyond our possibilities for participation. The first he called the primary world, the real world. The world of science was described as 'secondary' because it is derived from the first world. Problems arise when the latter is forgotten, when people think that the world described by science is the world as it really is, that humanity is in our genes, etc. Less explicitly, the same topic had already been mentioned in the opening lecture by Klaus-Michael Meyer-Abich, when he compared Goethe's way of doing 'participatory' science with conventional natural science. In Susan Lindee's talk it appeared as the distinction between DNA as a cultural icon functioning in the human life-world and the scientific DNA-concept.

The discussions in the plenary session focused on the relation between the scientific world and the primary world. One scientist considered this dualistic view of two worlds to be a dangerous myth. Scientists are not just detached onlookers, they also do experiments and every experiment is in some sense a form of participation with nature. Looking back on this discussion it appeared to us that a misunderstanding may have arisen here because of different ways of using the word 'participation'; participation in the sense of inner participation (communication) of the subject with the object of research, versus participation as physical interaction with the object of research. Van der Wal's distinction was not referring to a dualistic opposition of two worlds, but rather to a hierarchical relation between worldviews, different mentalities, which can be present in one and the same person.

There is a strong normative tendency in the methodology of natural science to rule out the subject as much as possible, and to describe/explain the outside world as a material object. The chairman gave two examples of this. A normative/ ideological example is Lewis Wolpert's book 'The Unnatural Nature of Science'. Wolpert says bluntly that doing science requires one to remove oneself from one's personal experience. Scientists should distrust 'natural' thinking, thinking that is rooted in our ordinary, day-to-day common sense. From Wolpert's book one could get the impression that he thinks that the world of science is the primary world, and the world of common sense the secondary world. This can be compared with the traditional distinction between the primary qualities of reality and the secondary ones, such as the experience of colours. Wolpert says: 'Physics teaches us that the greenness of grass, the hardness of stones and the coldness of snow are not the greenness, hardness and coldness that we know in our experience, but something very different'. Combined with another statement that it is a question of either scientific understanding of the world or dogma and ignorance, one can conclude that for Wolpert there are no alternative ways of getting knowledge about the world. The other example refers to empirical research about scientists doing animal experiments in the laboratory. Sociologists have described in detail the transformation of the 'natural animal' (the animal as experienced in our life-world) into the 'analytic

animal' (the animal as an object of scientific research). The experiential animal gets a name, the analytic animal gets a number. The same sociologists have also shown, however, that many scientists have emotional difficulties with totally objectifying animals – it is a very ambiguous relationship. This relates to the experience of one participant in the audience, that many doctors either depersonalize their patients, or they have a very schizophrenic relationship to them. At one moment they treat patients as human individuals and at another moment as medical objects. It may be the case that the process of objectivisation becomes more difficult as our object of research moves from animals to the human being. The same ambiguity was also mentioned in the lecture by Ernst Peter Fischer. He pointed out that attempts to replace the term gene by more exact terms failed, because of the pleasantness of the term 'gene'. Science is done by people with a soul; it is very hard to separate thinking from one's feelings completely.
From what was said during the conference we can conclude that Wolpert's view of two totally different worlds contains an element of truth as far as the scientific method is concerned.
But it cannot be true when we look at scientists as human beings, working in social institutions. It seems not possible and not desirable to bring about a total separation between two worlds, or to talk about science in the way Wolpert is doing. When scientists and science are seen as a part of the life-world the whole picture changes. Science then becomes a specific way of observing and thinking about the world. That science has become a dominant cultural resource in society became very clear in Susan Lindee's talk about the gene as a 'cultural icon'. She made clear that geneticists and molecular biologists play an important role in the establishment of the popular concept of the gene; there is much interaction going on between the two worlds. Another way in which natural science plays a role is through technology, which transforms the life-world in many ways. This relates to the question from one participant about the position of technology in the scheme of the two worlds. He did not feel happy with the idea that technology is something negative, belonging to the second (scientific) world, and estranging us from the first world of immediate experience. He gave an example of severely handicapped people who, by means of technology, could participate again in the world. With respect to this point we would like to comment that the use of technical artifacts in human life is much older than natural science. However, at the same time it can be argued that there is a very close relation between the objectifying, mechanistic and experimental methods of natural science and the modern technology based on it. The estrangement from the life-world may be related to the search for 'universal' natural laws of a high level of abstraction. This is what Jaap van der Wal meant when he said that modern technology shares the onlooker's mentality which is characteristic of modern science. Combined with modern industry (multinational corporations operating worldwide), a technology based on this kind of science has a strong tendency to become universal and to replace local technologies, technologies adapted to a

particular place and people. What has been said about technology in general has implications for the discussion about the relation between biotechnology based on molecular biological knowledge and techniques, and biodynamic agriculture. In biodynamic (organic, ecological) agriculture, discussions are going on as to whether modern biotechnology could and should be integrated within this form of agriculture. The present-day resistance to biotechnology cannot be understood without taking into account the differences in world views (basic attitudes towards nature and science) and social position of traditional and alternative forms of agriculture. These differences are closely related to the discussion about the two worlds. The question is, whether an alternative kind of 'science' (and technology) is possible which, in its methodology, is much closer to the primary world, both in its relation to people and in its relation to nature. In the plenary discussions the question whether a participatory biotechnology is possible, was continually present in the background. It seems to us that there is no room for a second (or third, fourth) kind of science in Wolpert's (and many other scientists') view. As long as this attitude towards alternative science (agriculture, medicine) reigns among scientists it will be very difficult for them to reach speaking terms. As has been said in the introduction, absolutist claims stand in the way of a real dialogue.

A similar issue came up in the discussions on the field of medicine. Due to increasing specialisation the distance between the real patient in her/his primary world and the scientist studying the causes of a disease seems to be increasing. There is a tremendous distance between the study of disease at a DNA-level and the real-life situation. The gap between a physician, who looks at disease as an element in the biography of a patient, and the geneticist who studies the same disease in the laboratory, is so big that it has become difficult to bridge it. One person suggested that it is important that there be more contacts between patients and geneticists. Scientists may then also be confronted with handicapped people who accept their quality of life, and who are against the idea that their existence might have been prevented by abortion. This shows that advice in relation to prenatal diagnosis should not be based on genetic information only. One scientist said that geneticists do not live in an ivory tower; there is a lot of pressure on scientists by patient organisations. Many people who ask for genetic screening already have a genetically handicapped child; we may assume that they know what it means to have such a child. We would like to comment that contact between geneticists and patients surely may be of help, but the methodological gap remains the same as long as disease is only defined at the molecular level, as if the 'real' cause can only be found at this level.

With respect to the two worlds, Ernst Peter Fischer said that they are complementary. It is perfectly possible for a scientist to be a scientist in the lab and an artist at home. Such a scientist goes from one world to the other. Probably nobody would deny this, but here again the more important question would be, whether a kind of science is possible in which the scientific and artistic element are combined. Several participants from the Goetheanum said that this is possible, and that Goethe showed how to do this.

Risks of genetic engineering

Of particular importance in Guenther Stotzky's talk was the difference between the behaviour of organisms in the laboratory (where 'it may work well theoretically') and the behaviour in an ecological system (the real-life situation). As a consequence of this, non-scientists, or scientists other than molecular biologists, have to participate in risk perception, risk assessment and risk management; it cannot be left to the molecular biologists themselves. Mae-Wan Ho put it more strongly by her emphasis on the unpredictability of the effects of genetic engineering and of bringing genetically modified organisms into the environment. Genetic engineering is based on the belief in the constancy of DNA, but according to what she called 'the new genetics' this is an illusion; the genome is dynamic and flexible. Johannes Wirz pointed out that there is more and more evidence that besides DNA-centred inheritance there exists an organism-centred inheritance with the environment influencing the genome. All the speakers emphasized the 'contextuality' of the genes and the unpredictabilities associated with genetically engineering living organisms. These lectures raised two important issues. The first was whether these uncertainties surrounding genetic engineering would not be a reason to press for a moratorium on genetic engineering. The second issue was whether, given the uncertainties, we really do need these new technologies; are there no alternative ways? Guenther Stotzky was not in favour of a moratorium on molecular biological research. Such a decision can only be justified when there is real evidence of risks. If we decide on a moratorium because of lack of knowledge about the effects then we also rule out the benefits we can have from applying these techniques. However, he saw the need for a moratorium on the commercial release of some transgenic organisms. Mae-Wan Ho pointed out that application of the precautionary principle makes it possible to defend a call for a moratorium even if we lack the knowledge for predicting the consequences. She is in favour of a moratorium on the commercial release of transgenic organisms. There is evidence enough and for most products envisaged there is no indication that we really want, let alone need them. Stotzky agreed that in discussions about risks not enough attention has been paid to the question of alternatives. Given the inherent uncertainty about risks, the question was raised as to how we know that we have done enough research in the laboratory before we go to the field. Even with a strict case-by-case approach we have very little knowledge about interactions within the field. Stotzky said that risk is a part of life, we cannot exclude all risks and we are willing to take some risks when there are also benefits to compensate for the risks. One participant compared the introduction of novel genes into the environment with the introduction of a tractor in a third world country. It is sometimes forgotten that a tractor can only function when petrol and spare parts are available. It is an end-of-pipe solution. It is much more important to start at the other end, in the real world so to say, and look for solutions at a regional or local level. Instead of looking at an insect pest as a plague which has to be eradicated by pesticides, one can also see it as an indication that something is wrong with the agricultural methods used. This way of looking generates a participatory technology. Mae-Wan Ho agreed with this. We have to look at life-cycles as the basis of sustainability; modern agriculture based on traditional scientific thinking has broken up these cycles, thereby causing problems.

Values and ethics

With hindsight we were surprised that the question of the two worlds also played such an important role in the plenary session about the risks of genetic engineering. In the methodological process of objectivisation and reduction many aspects of primary world of experience are left out. One of these aspects is values. The split between the world of facts (studied by science) and the world of values (studied by ethics) has already been mentioned in the lecture by Jaap van der Wal. But developments in the field of genetic engineering have led to an upsurge of interest in ethics. We are now seeing the formation of ethical committees, on local, national and international levels. The importance of ethics in education is emphasized in discussions and the Dutch Biotechnological Society has even accepted a professional moral code based on the precautionary principle. This is why the chairman brought in the question of the origin of our values. If science is value-free, where do scientists get there values from? What is the role of religious and other world views in this connection? Do they just belong to the private domain of our subjective feelings, outside the cultural and political domain of ethical judgment? Research into the processes of value-formation has shown that cognitive, affective and volitional aspects are involved. There is the (thinking) element of choosing freely from alternatives after consideration of the consequences of each alternative. Then there is the element of feeling happy with the choice made, being open to one's inner experiences and being willing to share these feelings and experiences with others. Finally the will gets involved when we ask the question whether and how often we really do something with these choices in our daily life. Values define what people think is the morally good life. One participant pointed out that not only acting but also refraining from an action (such as not using genetic engineering) can be morally relevant. This was an implicit endorsement of the importance of alternatives in ethical judgement.

Genetic engineering sometimes causes strong emotional responses and feelings, such as to the creation of *Drosophila* with eyes at different parts of the body, feelings which seem comparable with those raised by the 'mouse with the human ear' (which is not a result of genetic engineering). One person said that feelings are primarily personal and difficult to share with others. A distinction was made between emotional responses which are highly individual and true feelings which might be more objective. Do these emotions/feelings have any meaning for the public domain? Could they be indicators of some truth about values? Can they be raised to a more general level? One scientist pointed out that feelings are the driving force behind the public discussion. A representative of an animal protection movement said that feelings are not respected by scientists, they are ridiculed by them and seen as a sign of ignorance. This brought up the question of the role of feelings in the process of knowing.

It is usually recognized that perception and thinking are the basic elements of the process of knowing. One participant mentioned that the feeling of pain is usually taken to be a reliable guide in medical diagnostics. Western culture has difficulties in dealing with feelings because they are not accepted in the dominant tradition of natural science. A nurse endorsed this view, saying that in her surroundings feelings played a very subordinate role. This was no answer however to the question whether feelings could play a role in the acquisition of knowledge. One person said that there seems to be a relationship between feeling and intuition in this context. Also Goethe's phenomenological method was mentioned as containing an element of feeling in it, which is lacking in traditional science. There seems to be a link between feeling and the perception of wholes. A holistic way of perception may be of help in developing criteria for speaking about the 'integrity' of an organism. It became clear from this discussion that there are different levels of feeling and that feeling in the sense of being fully open to what is coming to us in (holistic) perception may somehow be important, not only in the process of knowing the world, but also in relation to the question of the origin of our values.

Eugenics

At first sight it might seem unreasonable to use the concept of eugenics as a heading for what happened on the last morning. This word has a very negative connotation. In using this heading we do not intend to imply that a eugenic motive was somehow present in anything that was said by the speakers in their lectures. The theme of eugenics came up several times in the plenary discussion and to some extent it closes the circle to the first speaker of the conference. Looking at the popular culture of DNA-thinking (genetic essentialism), as described by Susan Lindee and Dorothy Nelkin in 'The DNA Mystique', one gets a strong feeling that it is only a small step to a new form of eugenics. This new eugenics will not be imposed by the state for ideological reasons as in Nazi Germany, but will manifest itself from the bottom upwards. Independent of the motives of individual researchers, DNA-thinking may lead to the idea that there is a direct relation between (changes in) particular genes and (changes in) particular physical properties. When this knowledge is purposefully used to bring about changes in a human population by regulating the distribution of 'good' properties or eliminating 'bad' properties we have to speak about positive and negative eugenics. With eugenics we are used to thinking of state-imposed measures such as enforced sterilisation to bring about the desired goal. For the future it is possible to imagine a gradual development from free individual choice to abort a foetus with a 'defective genome', to social pressure on individuals to do this. Then it may only be a small step to state-imposed pressure not to bring a child into the world with a genetic defect at the (financial) cost of society. If, in the worst scenario, such practices become widespread (with genetic tests easily available), the effect could be the same as that of classical negative eugenics.

In his morning lecture Hansjakob Müller explained the shift in medicine from emphasis on environmental causes of disease to genetic dispositions being responsible for disease. He emphasized that genetic testing should not be done without genetic counselling, which should be based on the informed consent of the patient. The patient should be absolutely free (autonomous) in deciding to accept or not to accept genetic information or medical action based on this information. We would like to comment that in the light of the worst scenario of eugenics mentioned above, it seems of the utmost importance that this right of self-determination is protected by law. In the 'Convention on human rights and biomedicine', which has been accepted by the Ministers of the Council of Europe (November 1996), 'any form of discrimination against a person on grounds of his or her genetic heritage is prohibited' (article 11). Someone in the audience said in connection with the danger of eugenics that it is very important to discuss the question as to who decides about what is good, normal or abnormal. Another person mentioned the idea of 'a life not worth living' in this connection. In a situation of economic depression this idea seems to get more emphasis. In screening programmes for Down's syndrome in England there has been a lot of talk about 'cost-effectiveness'. This influences the attitude of physicians; a test is not done when the intention to abort a defective foetus is absent.

From Michaela Glöckler's lecture it became clear how much the idea of a life not worth living is related to one's conception of the human being. Against the idea of genetic determinism (essentialism) she put forward the idea of the individual being 'who plays the piano'. She took up the theme of the lectures of the second day about the fluidity of the genome, and its openness to environmental influences. She underlined the influence of a third factor besides genes and environment, the human individual mind, the true self, which cannot be separated from the body when incarnated in this body. From this perspective disease may also be seen as helpful to individual development. Whether a life is worth living can only be decided by the individual (the true self), it cannot be decided by others. In answering questions from the audience she said that she could not imagine how life could exist without pain. It also seems to be important to see pain in the social context, the feelings of care unbraced by illness, etc. Others in the audience warned against the 'idealisation' of handicaps; some patient organisations have rejected the view of pain or disease as a positive factor in one's development. Koos Jaspers said in his lecture that there will be more predictive tests in the future because people want them. To find out the function of genes they are knocked out in mice. Jaspers expected that in future experiments will also be carried out on human pre-embryos, because of the differences between mouse and man. He was very much concerned about the question as to what the effects will be on the self-fulfilment of people. Early diagnosis may have the effect that disease will already be experienced in the mind long before it actually occurs – if it occurs at all. Questions were asked about the pre-embryo. Is it a good model for the adult human being? Jaspers emphasised that the pre-embryo concept refers to the stage of embryological development before implantation (attachment to the uterus). This stage was distinguished to make research on this human 'material' possible; it is called material because it is not yet considered to be an individual person (twins can still arise in this stage). It is a better model than a mouse. One person said that only two percent of all diseases can really be said to be 'genetically determined'. Most diseases are developmental diseases; even knowing the genetic factors involved it will be difficult to give predictions for the future. The idea that diseases should be eradicated would come down to the elimination of all flexibility and adaptability, and thereby the elimination of life itself. One participant finally asked what we have learned from this conference? What are the consequences for the everyday life of young scientists working in the lab? This question was not answered, and the conference was not intended to provide clear cut answers to this question. Many participants felt that the discussion climate during the plenary sessions really came close to that of a power free dialogue. As has been said in the beginning, *Ifgene's* main task is at the cultural level, in the hope that what is happening at this level will inspire individual people in their individual decisions in everyday life-situations, or in political life.

13 Gene concepts in motion: from Mendel to molecules

WORKSHOP

JACQUELINE GIRARD-BASCOU
Biologist
Institut de Biologie Physico-chimique
13 rue Pierre et Marie Curie, F–75005 Paris

CRAIG HOLDREGE
Biologist
Hawthorne Valley School
169 Route 21C, Ghent, NY, USA

1 The story of the gene from Mendel to molecules: a classical drama in five acts

JACQUELINE GIRARD-BASCOU

In the prologue, the gene concept is a rumour.
DNA is sleeping in the nucleus as the sleeping beauty in her castle. From the beginning, no one has been able to disturb DNA. But *a rumour went abroad in all the country of the beautiful sleeping Rosamond.* This rumour told us that somewhere, something in the seeds or in the blood has the power to maintain the heredity of characteristics.

The first act begins with Gregor Mendel and hereditary factors (1865).
Gregor Mendel lived in the last century in Czechoslovakia. He was interested by hybridization between ornamental plants. Mendel chose three conditions for his experiments. First, to study alternative constant character traits, for example, yellow or green seeds, red or white flowers. Second, to have the possibility to hybridize without contamination by pollen of another plant. Third, to obtain fertile progeny. The chosen plant was the common pea (Pisum sativum). He crossed plants differing in one or two traits and he observed the resulting hybrids in the first and second generation. In the first generation, hybrids had the traits of only one parent but in the second generation, the two parental traits reappeared in constant proportions.

From these experiments, he postulated new concepts and laws which were later developed in a more explicit form. He proposed: 1) The existence of *hereditary factors* that are responsible for traits; for example factor for the colour of seeds or factor for the colour of flowers. They were later called *genes.* 2) Each hereditary factor exists in alternative forms. They were later called *alleles.* 3) Each form exclude the other form in the gametic cells. As a consequence, each gamete contains only one factor and each plant resulting from a fertilization contains two factors. This alternation one factor/two factor in a reproductive cycle was later described as the alternation between *haploid and diploid phases.* 4) In a hybrid, one form is dominant, on the other form recessive. The concepts of *genotype and phenotype* were thus introduced.

The second act progresses with Thomas Morgan and the chromosome theory of heredity (1915).
The Mendelian laws were rediscovered at the beginning of our century. It was shown that they can be applied to certain traits of plants, animals and human. Morgan chose to work as Mendel. He chose to work with the prolific fruit fly, *Drosophila.* He crossed flies of strains differing in one or several traits. At the beginning, he used spontaneous mutants. But later, he obtained new traits by mutagenesis with chemical compounds or radiation, which were known to affect DNA. He chose a wild type strain constant for all its traits and by mutagenesis treatment he obtained a collection of mutant strains. For example, the wild type flies had red eyes, mutant flies can have white eyes. This modification of the experimental approach has consequences for the language used. With Mendel, the hereditary factors were called by the character, for example the factor for the colour of flower. With Morgan, the genes were called by alternative mutated traits. For example, he called *white* the gene which by mutation gave white eyes. But the wild type form is red eyes. By calling genes by the mutated trait, which in most case corresponds to an inactive form of the gene rather than to its wild type functional form, the genetic language is inverse in respect to the function of the gene. This situation is even clearer with *lethal* genes, which by mutation lead to the death of the flies. It is evident that these genes in the wild type are necessary for life, they are *vital* genes.
Morgan and coworkers confirmed Mendel laws and concluded that the genes are arranged linearly and form linkage groups. At the same time, the cytologists observed linear structures in the nucleus: the chromosomes. They showed that each gamete contains a set of different chromosomes, the plants or animals contain two sets of chromosomes. Thus the distribution of chromosomes in gametes and embryo appeared similar to the distribution of the postulated genes. This convergence suggested that chromosomes can represent linkage groups. The result was the chromosome theory of heredity in 1915. The chromosome could be described as rosaries of genes.

In the third act a dogma is established with Watson and Crick and the structure of DNA (1953).
During a period of thirty to forty years, research was intensified and focused on bacteria, viruses and fungi. It was the period of microbiologists and biochemists. A lot of conclusions were drawn. First, the gene is constituted of DNA, a substance found in chromosomes. Second, DNA is a double helix of polynucleotides. Each nucleotide contains the same sugar, a phosphate group and a nitrogenous base which belong to two chemical families: purines (adenine and guanine) and pyrimidines (thymine and cytosine). They were commonly named A, G, T, C. In the double helix in front of an A there is always a T and in front of a G there is always a C. Watson and Crick received the Nobel Price for the discovery of the structure of DNA in 1953. Third, a sequence of DNA is transcribed into a sequence of nucleotides called messenger RNA. In this sequence, the sugar is modified and the nitrogenous base T is replaced by the nitrogenous base uridine (U). The messenger RNA is transferred in the cytoplasm and translated into a sequence of amino-acids which correspond (in most cases) to a protein. A combination of three nucleotides is necessary to code for one amino-acid. Thus a gene became identified as a sequence of DNA that was transcribed into the RNA coding for a single protein. Jacques Monod who received the Nobel Price with François Jacob in 1961 for a model of regulation of the expression of bacterial genes, claimed that "*what is true for bacteria is true for an elephant*"!

In the forth act, the drama is getting form at the conference of Asilomar in 1975.
Doubts and fears became to be perceived in the scientific community. At the conference of Asilomar (US) in 1975, a group of renowned researchers decided to place a moratorium of two years on genetic experiments to evaluate their dangers and to decide the rules of security to apply. These fears had two origins. First, most of the genetic experiments were done with a bacterium, *Escherichia coli*, which is commonly found in the human digestive tract but when they are in another part of the body they are a cause of diseases. Very often, for the easiness of experiments, the manipulated bacteria are resistant to antibiotics or carry part of viral genome. The risk was that these manipulated bacteria could transfer genes for antibiotic resistance or other uncontrolled genes to bacteria of our normal life. Second, the new results indicate that a gene is not always a continuous, well defined sequence of DNA coding for an amino-acid sequence; the reality is more complex than thought. For example, most of the eucaryotic genes are interrupted, they are constituted of coding parts (exons) and non-coding parts (introns), these introns are removed from the RNA molecules transcribed from the DNA by a splicing mechanism; furthermore, upstream and downstream of the coding sequences, sequences are found that are necessary for the expression of the gene, thus the ends of a gene became difficult to determine. It was also found that some genes overlap. The dogma elaborated in the third act cannot be maintained.

Table 1

Drama	Media event	Gene concept
Prologue		The gene concept is a rumour.
Acta I	Laws of heredity Mendel (1865)	One hereditary factor = one character trait. (alternative forms) = (white or red flowers).
Acta II	Chromosome theory of heredity Morgan (1915)	Chromosomes viewed as rosaries of gene. Induced mutations by radiations or chemical compounds. Genes were named by mutated traits.
Acta III	Structure of DNA Watson and Crick (1953)	One gene (sequence of DNA) = one protein (sequence of amino-acids). One structural model: linear relation with one sequence of DNA, one messenger RNA, one sequence of amino-acids.
Acta IV	Moratorium Conference of Asilomar (1975)	Several structural models: interrupted genes, overlapping genes, no definite ends
Acta V	Human genome project (1988)	A gene cannot be defined as one sequence of DNA. A gene is an hereditary information for an elementary function in the cell that is stored in DNA.

Acta five: the human genome project (1988)

Now, with the new technologies, it is possible to manipulate genomes and also to sequence entire genomes. The genomes of several bacteria and recently of yeast (3×10^7 nucleotides) have been completely sequenced. The international project to sequence the human genome (3×10^9 nucleotides) began in 1988. For about ten years, the main aim has been to find the genetic anomalies that are responsible of human diseases. Most of experiments are now done with higher organisms and humans. Again new results brought surprises: 1) The same sequence of DNA can give several messenger RNAs by alternative splicing and thus one sequence of DNA can correspond several proteins; 2) For some genes in organelles, the sequence of DNA has gaps, nucleotides must be added after the transmission on the RNA molecule by a mechanism called editing; 3) Most of chromosomal DNA of higher organisms does not code for proteins or RNAs; 4) Genomes are more fluid than we thought. In human, more than 30% of the DNA are repeated sequences that can vary greatly in number from one organism to another (these sequences can be used in forensic medicine to identify a person). About 10% of the genomes are constituted by sequences that can move by transposition, provoking rearrangements.

Thus it was no longer possible to define a gene as one sequence of DNA. A gene is better defined by its messenger RNA, but the latter is a short-lived molecule. Now we define a gene as an hereditary information for one elementary function in the cell that is stored in DNA.

II Plasticity in Human Heredity

CRAIG HOLDREGE

In her introduction, Jacqueline Girard-Bascou vividly described the way the concept of the gene has changed from Mendel to the present. Let us now move to the world of phenomena that genetic theories and models try to explain. I will describe phenomena that point to the plasticity that lies at the basis of all heredity. We often forget this dimension when focusing on genes and traits. Moreover, these phenomena show that human development is embedded in a rich and multifaceted context. It becomes evident that it is a grave oversimplification to think of genes determining the characteristics of an organism.

If we were not born in so helpless and germinal a state, if we did not then go through an extended (actually life long) period of development of physical and mental capacities, which are subject to myriad internal and external influences, if we did not also mould our own lives struggling to free ourselves from hereditary and environmental determination – in short, if we were not human – it would be much simpler to speak about heredity in the human being. As it is, there is no characteristic that could be simply defined as purely hereditary as opposed to environmentally conditioned.

The feet and legs in the context development

An infant often becomes aware of its feet when it bites into its toes for the first time. These organs remain fairly dormant until the child, approximately around its first birthday, pulls itself up and stands – or better, wobbles – upright. With this act the child begins to take hold of its whole body. If one studies the feet and lower limbs, one sees that they are at first by no means adequate instruments for standing and walking.

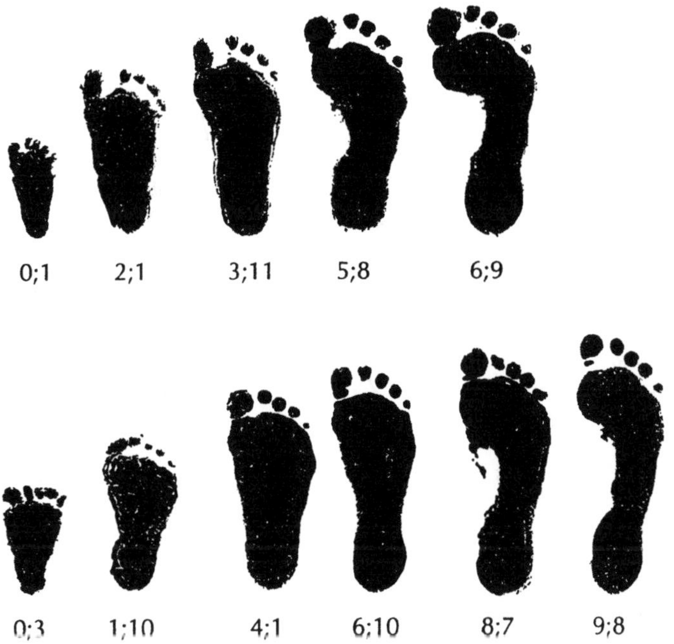

Figure 1.
Development of the arches in two different children (sibling girls); ages are given in years and months (from Holdrege, 1996).

Figure 1 shows the foot prints of two different children. The arches are not very defined until about four years of age or later. In the one child the arches are quite pronounced by seven, in the other child by eight years and six months. In the time between the second year of life, when a child begins to stand and walk, and the age of seven or eight, when its arches are essentially developed, the child has also learned to jump, run and skip. This means that its feet develop in the context of their usage. We do not first have highly developed feet and legs that are ideal for standing, walking, jumping, running and skipping. Rather, we stand, walk, jump, run, and skip with unfinished instruments that develop in this process. *The feet we walk on when we are seven are not ones we have inherited.*

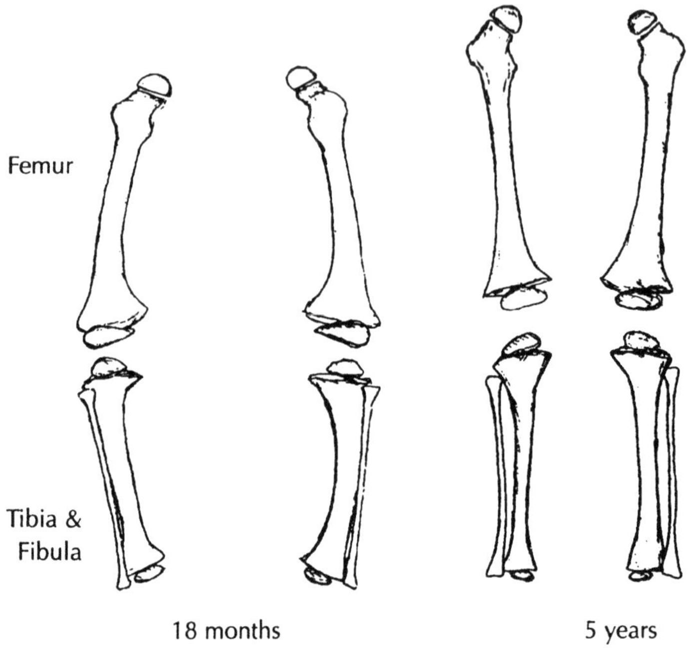

Figure 2.

Development of the leg bones of a boy between the ages of eighteen months and five years. Notice the remarkable straightening of the leg bones. The disks at the ends of the bones-epiphyses-fuse with the shafts only after longitudinal growth ceases (around twenty years of age). (Author's drawings after x-ray pictures from Holt et al., 1954; reprinted with permission from Holdrege, 1996.)

Figure 2 shows the leg bones of an eighteen month old boy and the same boy at five years. The transformation is remarkable. At eighteen months the boy shows a so-called physiological bowing of the legs (Holt et al., 1954). Most children who begin to walk have bowed legs, but not in such an extreme fashion. The lower ends of the femurs and the upper ends of the tibias show strong flaring on their inner (medial) surfaces. The boy already had bowed legs when he began to walk at eight months. Such bowing increases the compression and stress on the medial surfaces of the bones. The bones answer to this condition and the medial flaring of the leg bones at the knee joint increases.

Now, if the bones had continued to accommodate the stressing in this manner, the flaring would have worsened, the bones would become heavier and more unwieldy, but at the same time the whole structure would be more stable. The legs would become somewhat ungainly columns to stand on, and difficult to move with. Instead, the bones straightened and the flaring decreased, as the illustration shows. The legs became relatively light and straight columns. They lent stability, but also made movement possible.

In their form the legs follow the upright posture that the child has taken on and been exercising for over three years. This indicates that not merely an accommodation to stresses is occurring, but in a certain sense stresses are being overcome by the tendency toward uprightness. We gain uprightness by counteracting what weighs us down. There is no uprightness in a world of weightlessness. The weight of the body both resists and stimulates our becoming upright. And in this interaction our body is formed.

Erwin Straus writes: "Upright posture characterizes the human species. Nevertheless, each individual has to struggle in order to make it really his own. Man has to become what he is ... While the heart continues to beat from its fetal beginning to death without our active intervention and while breathing neither demands nor tolerates our voluntary interference beyond narrow limits, upright posture remains a task throughout our lives ... In getting up, in reaching the upright posture, man must oppose the forces of gravity. It seems to be his nature to oppose nature in its impersonal, fundamental aspects with natural means. However, gravity is never fully overcome; upright posture always maintains its character of counteraction. It calls for our activity and attention." (Straus, 1966, p. 141)

This counteraction is present from the moment a child initially brings itself to stand. We live in this wilful process, overcoming the resistance of gravitational and mechanical forces, which also include the resistance of the not yet ideally formed body. The body is then being continually moulded and remoulded and becomes a fine instrument for the activities which we carry out. It adapts to the way we stand and move in the world, it becomes a reflection of the way we exist in the world. Our body becomes ours.

We can thus begin to look through the body, as it were, towards the individual. Imagine a young boy who, from the time he begins to stand and walk, always stands and walks on tiptoes. He continues to live in his body this way and as an adult still usually walks and stands on tiptoes. Already as a child he had unusually large calf muscles and the front part of his feet were very wide. His body was forming in accordance with his activity. It is a riddle why a child takes on such a posture and gait. It results in a unique way of relating to the body and the world, but is also a reflection of the will (I do not mean here a conscious process) to do so. If I awaken to the riddle of this phenomenon, then it points beyond itself to the unique individuality of this person.

References

Goethe, J.W. (1988) *Scientific Studies*, Ed. Douglas Miller, Suhrkamp, New York.

Holdrege, C. (1996) *Genetics and the Manipulation of Life: The Forgotten Factor of Context*, Lindisfarne, Hudson, NY.

Holt, J.F. et al. (1954) Physiological bowing of the legs in young children, *Journal of the American Medical Assoc.* **154**, pp. 390-394.

Love, J. (1991) Animals in research. *Nature* **353**, p. 788.

McKusick, V.A. (1986) *Mendelian Inheritance in Man*, John Hopkins University Press, Baltimore.

O'Neill, L. et al. (1994) What Are We? Where Did We Come From? Where Are We Going? *Science* **263**, pp. 181-183.

Straus E. (1966) *Phenomenological Psychology*, pp. 137-165. Basic Books, New York.

14 DNA and food technology – between natural food and food design

WORKSHOP

ANNEKE HAMSTRA
Household and consumer researcher
SWOKA
Schipholweg 13-18, NL-2316 XB Leiden

MIRJAM MATZE
Nutritional researcher
Louis Bolk Institute
Hoofdstraat 24, NL-3972 LA Driebergen

Abstract
In this workshop the participants were invited to find out by themselves some different viewpoints existing exist in discussions about quality of food and food production. By making them role players they will discover very basic differences between fundamental attitudes towards nature and health.
In a more theoretical way the philosophical aspects of four fundamental attitudes are explained. Debates between producers, consumers, farmers and environmental groups often end in an impasse. This fundamental attitudes can be used as a handle or an instrument in understanding each other and forming judgements.

Fundamental attitudes, DNA and food production

Introduction

Although our workshop, on food production, was cancelled in favour of others, we would like to give you an impression about our intentions.
Discussions about food and food production often leave an unsatisfactory feeling. Between various groups, e.g. consumer organisations, farmers, food processing industry

and environmental groups, the debates often end in an impasse. This is most visible in the discussions about gene technology in food production.
The debate is often about various aspects, like safety or health effects for humans, environmental effects, and economic necessity. In fact, the basic question: do we want foodstuffs, made with genetic engineering, on the market or not, may remain underneath the general level of the discussion.
Apart from the discussions on all kinds of rational arguments on safety etc, the discussions have also an emotional aspect. What is the meaning of this?
Our statement is that people are often unaware of the fact that their ideas and opinions are fed by the way they view the world. How we view the world, and mankind, define our attitudes towards scientific and social questions and developments. When the basic world views, or fundamental attitudes (towards nature, mankind and technology) differ, it is difficult to understand each other in a debate. As these fundamental differences remain unspoken, implicit, the debate may easily end in an impasse and in unsatisfactory feelings, because the same arguments may have a different meaning and value for each of us.

Introduction of fundamental attitudes
In 1993, the philosopher Kockelkoren wrote a report on ethical aspects of biotechnology in plants, for the Dutch Ministry of Agriculture, Nature Management and Fisheries. In this report, Kockelkoren re-introduced the concepts of Zweers (1989), which can be called 'fundamental attitudes'.
These attitudes can provide more insight in the viewpoints of people and the consequences these viewpoints have for the way they view and deal with the world, in their personal life and in their work.
These attitudes can be seen as an instrument for a better understanding of the deadlocks in economical, agricultural and nutritional debates.
We feel that awareness of these fundamental attitudes, and awareness of our own attitude position and that of others, will help to be able to understand and respect other parties in the discussion. Mutual understanding and respect may subsequently lead to a better starting point to learn from each other and to face the consequences of these differences. Sometimes they can be overcome, sometimes acceptance parties can agree to disagree. In that case, pluralistic solutions can be more satisfactory than compromises.
For this conference, we have chosen the subject of biotechnology and food production. This moment, the first food products of genetic engineering are entering the market. In this workshop, we would like to discuss the various fundamental attitudes that play a role for all those people who have to deal with products of biotechnology: technologists, researchers, farmers, consumers, physicians, sociologists and politicians.

Workshop setup

Instead of starting from the philosophical background, we thought it would be interesting to start talking about food products. What would, for example, be your ideal potato? Which aspects are important for a good potato? Of course, this would be different for a consumer, a potato grower or a retailer. Within the group of potato growers, there will be differences in view between an organic farmer and an 'ordinary' farmer. Within the group of consumers, we would expect differences between consumers who like convenience, or those who give priority to health. For retailers, there is of course a difference between a manager of a large supermarket and, for example, a health shop keeper.

By asking six of the workshop participants to assume a role (e.g. convenience consumer, health shop keeper), each would come up with a somewhat different set of important criteria for a good potato. After an inventory round (which aspects are mentioned), the role players question each other to see what exactly is meant by each demand or aspect. For example, if someone says 'it should taste good', he or she is asked to describe more in detail what this taste should be.

Meanwhile, the other participants would have been asked to observe the discussion and to note aspects which express conflicting demands between the role players. For example: 'natural' on the one hand; 'constant quality' and 'high yield' on the other.
After this, a second round with other role players and other roles would be done with another product (a frozen meal).
The result would be that the participants made an inventory of (conflicting) aspects. These were noted on stick-on pieces of paper, and could then be placed on a large sheet of paper, along one or two dimensions. Aspects that conflict very much would end up at the extremes of the dimension. Aspects that could go very well together, would be placed at the same side of the dimension.
Although we could of course not predict the exact results, we would expect that the two-dimensional field would show groups of aspects and criteria for products which would relate to a certain fundamental attitude. We would expect to find a group of aspects which would represent an attitude towards an instrumental world view: products which are tailor-made according to exact technical norms. These products are stable in quality, easy to process, efficiently produced and have a sharp price-quality relation. On the other hand we would expect to find a group of aspects relating to a more ecological view on nature, with more individual products with more variety in taste, higher price, and a more natural way of production.
The workshop would then continue with a presentation of the theory of the four fundamental attitudes.

Table 1. Agriculture, nutrition and technology: between money and mind

image of man	– machine	– living machine	– living organism	– spiritual being
health and nutrition	– food is fuel – calculable quantity of carbohydrates, proteins, fats, vitamins and minerals	– health-awareness on nutrientlevel: more vitamins, less fat, meat and sugar	– to care about vitality and self-regulation – homeopathy	– to stimulate vitality and spiritual development
	– collective needs	– collective needs	– individual needs	– individual needs
	– food design, functional foods	– varied food	– fresh and natural food	– health food
agriculture	– chemical agriculture	– environment-friendly agriculture	– organic agriculture	–bio-dynamic agriculture
economy	– as cheap as possible	– should be payable within conventional economic thinking	– more costly products accepted – willingness to break through the economical thinking	– more costly products accepted – ecological economy
	– economic norms	– environmental norms	– strict norms	– intrinsic norms
	– technology push	– technology drive	– technology management	– technology masterment
fundamental attitude	**dominator**	**steward**	**partner**	**participant**
image of nature	– earth is giving raw materials	– earth is giving raw materials	– nature is ally of man	– man is strictly interconnected with nature
	– functional without limits	– functional but not unlimited	– to respect the intrinsic value of nature	– to respect the intrinsic value of nature, not to attack them
	– inexhaustible	– accountability towards the Creator and towards future generations	– to support the selfregulation o nature	– to support the development of the earth as a organism
	– maximum profit and benefit	– growing environmental awareness of short-term solutions	– to keep the earth in balance	

Theory of the fundamental attitudes

Table 1 gives in a very schematic way an overview of the classification in four fundamental attitudes towards nature, mankind and technology. Most pregnant are the image of man and the image of nature. This are keywords in understanding how an individual look at the world.

If man is considered as a machine his food is the fuel and his heart is the pump. So food has a functional aim.

The endeavour is to control the world and to make man independent of climate, seasons and agricultural restrictions. In this makable world agriculture can be steered by using chemical substances. Food can be made by food design with the help of vegetable, animal or non-agricultural (bacteria, fungi) materials. Genetic engineering can be a great help in producing agricultural and nutritional products. This economy friendly view is borne by the so-called *dominator*.

The *steward* is looking to the human body as being it a biochemical system or a living machine.

It needs balanced biochemical content as: not to much cholesterol and saturated fat, not so much sugar ('light-products'), and ample fibres, vitamins and minerals. Personal health is a deciding factor in consumers choice of products, in addition to price, ease and efficiency.

Adapted technology can be a tool in this so-called environment friendly agriculture. Normalization is important in industry and agriculture to control environmental impact. Biotechnology is being used to combat unfavourable side effects. For example, modified bacteria can be used in the purification of polluted water. A steward is realizing his accountability towards the Creator and towards future generations.

From a nature friendly viewpoint man is regarded not just as a biochemical system, but as a living organism who is integrated in a wider setting. Man is as a *partner* related to nature and they are allies of each others.

Man has to care about his vitality and self-regulation. Even so man is recognizing and supporting the intrinsic value and self-regulation of nature. Organic agriculture is region and season dependent. Chemical pesticides and fertilizers are not to be used. Undesired side effects are not managed by using (bio)technological methods, but the frameworks of thought are readjusted to see the side effects as an expression of an overly unilateral approach to nature.

A healthy lifestyle and consumption of fresh, natural and little processed food are considered as beneficial for good health and for keeping the earth in balance.

From the *participant* point of view man is seen as a spiritual being and earth as a living organism.
Mankind, earth and each individual go through a developing process. Health is dependent of individual needs and lifestyle. The biography of each human is a keyword in the spiritual development. Nutrition is not only seen as supporting the vitality of the human organism, but also as supporting the development of the human being and the earth.
From this participant perspective the process of food production is of great importance. In biodynamic agriculture a farmer tries to establish a true-to-type growth process whereby the unique characteristics of each crop or each cattle-breeding can be manifested. Intrinsic norms of agriculture concern animals, but also plants. Real hesitation in applying biotechnology persists, since many people acknowledge the complexity and dynamics of processes between humans and nature, and recognize the still insufficient scientific orientation towards these processes in order to act in a morally responsible way.

Continuation of the workshop

We would then discuss firstly, whether the work of the participants has yielded a result which is recognizable from the theory. Secondly, it is interesting to discuss whether the understanding of the fundamental attitudes really makes the discussion clearer. To elaborate this further, small groups of participants would have been asked to describe and evaluate (form an opinion about) several products, starting from each of the fundamental attitudes. Product examples included e.g. coffee, milk with extra calcium, the genetically modified Flavr Savr tomato, irradiated mushrooms, cholesterol-free cheese made without animal substances, novel protein foods, lettuce grown in hydroponic culture. What would the four fundamental attitudes imply for these products? Is it from each point of view a desirable product, or not?

We would have hoped that the participants would have become more conscious of their own attitudes towards food production, and would have considered the four fundamental attitudes as a mean to explain conflicts: between different actors, or even within themselves. As organizers, we would have been very curious to see whether participants felt this approach was a contribution to their understanding, and a way forward for articulation and policy development in reaction to the public unease about the food production.

References

Hamstra, A.M. (1995) *Consumer acceptance of food biotechnology*, SWOKA, Leiden, The Netherlands.

Huber, M., Matze, M. and Lammerts van Bueren, E. (1996) Biotechnologie: vier impliciete visies op gezondheid, *Geno* 5, Louis Bolk Institute, Werkgroep Genenmanipulatie en oordeelsvorming, Driebergen, The Netherlands.

Kockelkoren, P.J.H. (1995) Ethical Aspects of Plant Biotechnology, Report for the Dutch Government Commission on Ethical Aspects of Biotechnology in Plants, in *Agriculture and Spirituality, Essays from the Crossroads Conference at Wageningen Agricultural University*, International Books, Utrecht, The Netherlands.

15 DNA and education

WORKSHOP

JOHN ARMSTRONG
Molecular Biologist
25126 Pleasant Hill
Corvallis, OR 97333, USA

FRANS OLOFSEN
Biologist
Geert Groote School
PO Box 77779, NL-1070 LJ Amsterdam

Teaching modelling in the context of genetic engineering

JOHN ARMSTRONG

Abstract
The construction of models is crucial to the practice of present-day science. Scientists view model-building as a essential means to simplify reality into understandable concepts about environment and all of its varied forms of life. Moreover, the molecular biologists and biotechnologists who manipulate DNA depend heavily on models to help visualize the molecules, both in the course of laboratory experiments and in the living organisms themselves. However, because modelling has such a central role in science, educators often do not address the limitations imposed by models. The purpose of this paper is to introduce some methods and ideas for teaching high school and college students about the advantages and disadvantages of modelling in the context of DNA and genetic engineering.

Traditionally models of chemicals have taken the form of balls and sticks to represent the atoms and chemical bonds. But in molecular biology model-building is more varied and complicated, often taking the form of computer generated images or descriptions of organisms as aggregations of biomolecules. Educators like to use such images and descriptions – these models of the DNA and its genes – teach molecular biology and biotechnology, since these visual and imaginary aids are thought to give students clearer pictures of molecules and organisms. Indeed to a great extent, the rapid pace of biotechnology is due expressly to the successful use of models. But with so much emphasis on their value, teachers can easily fail to remind students that these "pictures of reality" are mere, sketch-like representations. Only by giving much more attention to both the uses *and* the misuses of models can teachers at all levels avoid leading their students into a false view of reality.

Like all models, models of DNA must also inevitably be indirect representations – intellectual abstractions – the substances and their associated phenomena. To introduce students to a sound basis of how models are used by scientists in general – and genetic engineers in particular – the teacher can start with observations about model-building process in general. These observations can then be applied directly to examples from molecular biology and biotechnology.

One way to introduce the topic is to have students read Plato's *Republic* wherein he describes his famous "Allegory of the Cave" (Hamilton and Cairns, 1961, pp. 747-749). The class can discuss the relationship between the actual objects behind the people in the cave and what the people actually see on the wall as projected shadows of the objects. The students will quickly identify the shadows as models of the objects – simplified versions of the unseen objects that give only a partial view of what is hidden. To give a necessary balance between the advantages *and* disadvantages of modelling, the class should also discuss how the shadows do reveal some information, albeit limited, about the objects.

An excellent, concrete demonstration of Plato's allegory can be performed with two wire figures: a cube and a cylinder. Making the cube is straightforward, but the cylinder is constructed with two circles that are connected with two wires at the diameters in such a way that the figure looks like a square when viewed from the side. Shadows of these figures can be projected on a wall by means of a light from a slide projector, transparency projector or flashlight. The demonstration is begun by holding the cube so its shadow will be a square and turning the light on. Students discuss how the cube and its projected image are analogous to the objects and shadows in Plato's story. The students can pretend they do not see the cube and discuss how the square might be used to model what is unseen. What can be said about the cube from the square on the wall? What remains unsaid? Then the light is turned off, the cube is rotated 90 degrees and the light is turn back on. Again students can discuss how this second image of a square on the wall relates to the 3-dimensional figure.

The demonstration is repeated with the cylinder. But this time the wire figure is hidden from view inside a large box with holes on opposite sides to allow the light to pass through. First the cylinder is held to project a square. Again the students discuss what the wire figure might look like based on the shadow-model. The teacher can help lead the students to question the possible limitations of their descriptions of the unseen figure. The light is turned off and the figure rotated 90 degrees so it will project as a circle. The image of the circle should surprise many of the students and thereby emphasize the limitations of their initial conclusions. The students can draw possible structures for the wire figure and discuss what else they need to know to be certain their drawings are accurate. Finally they can discuss how their observations can relate to the uses and misuses of modelling in general.

To delve into models in the context of genetic engineering, students can read James Watson's *The Double Helix* and look specifically for his many descriptions of the role of modelling in the "discovery" of DNA's double helix structure, for example: "The α-helix had not been found by only staring at X-ray pictures; the essential trick, instead, was to ask which atoms like to sit next to each other. In place of pencil and paper, the main working tools were a set of molecular models superficially resembling the toys of preschool children ... All we had to do was to construct a set of molecular models and begin to play." (Watson, 1968, pp. 50-51)

Students can discuss selected passages like this one, both in the context of the utility of modelling and from the perspective of what might be lost by a simplified view of living organisms. The students can be encouraged to refer to the experiences of the people in Plato's cave and to the demonstration with the wire figures.

Since X-ray crystallograms are analogous to the shadows on the wall, another way to approach DNA models is via X-ray crystallography images of crystallized DNA (Watson, 1968, pp. 73 and 168). Of course the students will not know the methods of X-ray crystallographic analyses, but this will not deter them from recognizing some of the possible limitations imposed by studying these "projected" images of the DNA. On the other hand, the class will also be able to appreciate how the modelling was necessary to enable the scientists to make certain types of conclusions about the DNA – specifically, its physical-chemical characteristics. The teacher can also show the students pictures of a salt crystal, a protein molecule and a DNA helix and invite the class to discuss how the pictures may or may not relate to the X-ray crystallogram of DNA. If anyone says the X-ray crystallogram is DNA, the teacher can ask how the student knows this. This can lead into conversations about how people acquire their underlying biases about science from education, the media and other sources that present versions of what people know about the world as if these are the "true" reality. The class can also discuss the implications of this issue in the context of education in general.

There are numerous ways of depicting DNA and genes and describing heredity, that can be used to discuss modelling. For example, the students can be shown the base sequence of a gene and discuss it in terms of the genetic code as a model for the human being. What kinds of information about organisms can be gained from such sequences? What misinformation might people surmise about life if they study only the DNA sequences? In this context, the class can also discuss chromosome maps which indicate the locations of genes "known to cause" various "abnormal" conditions. How is the information on these maps like a model for illness? What are the possible uses and misuses of human chromosome maps as models of the human being? How might DNA sequences and chromosome maps lead people to overlook possible roles for a human soul and spirit? Other examples of topics that could be taught in a similar way include: the historical changes in the concept-model of "the gene"; gel electrophoresis as a method to construct a model of DNA and genes as manipulable fragments; the relationship between the sequences of human beings and other primates in the context of what it is to be a human being.

The "what it is to be human" theme leads directly into questions about the limitations of "reducing" people to models based on genes and playing down other factors that may determine who we are as human beings. Three books are available which enable a deeper study of how to approach this theme. One is *Mapping and Sequencing the Human Genome* (BSCS and AMA, 1992). Put out by the Human Genome Project, the book contains a series of exercises on the ethical, legal and social implications of gene therapy and emphasizes a view – a model – of humans as genetically determined beings. The second book, Holdrege's *Genetics and the Manipulation of Life: The Forgotten Factor of Context* (1996), offers a valuable perspective that points to the drawbacks of placing too much emphasis on DNA and genes in modelling human beings – in fact all living organisms – with concepts that take the organism "out of context." Holdrege does not address directly the subject of modelling. However, when he speaks of "context" he means the greater picture which is eliminated as soon as a simplified concept – a model – is used to describe organisms via genetics and molecular biology. Students could read this book and look at it specifically for what it implies about models. In their discussion, the students could focus on how modelling – even in the case of Mendel's early work with pea plants – has also spurred advances in the field. The class can point to ways in which such historical work influenced present-day views of the roles played by DNA, genes and heredity in the life and evolution of all organisms. As a contrast to Holdrege's "holistic" approach, students can read the views of molecular biologist Francis Crick in *Of Molecules and Men*, for example:
"Studying the cell as a living unit enables us to see in a broad way what it does. The system is usually too complicated to enable one to deduce from such experiments the *exact* details of the mechanism inside the cell." (Crick, 1966, p. 13)

Crick is quick to mention that "artifacts" are produced by this process and argues for a reassembly of the parts to understand the whole. Issues for discussion can focus on whether this process of breaking apart and reassembling – a combination of model-building and inductive reasoning – yields an accurate, total view of the original phenomenon. Other related questions for discussion could include:

- Does the manipulation of genes across the natural barriers between organisms affect the foundations of nature?
- Does the "DNA-thinking" of molecular geneticists influence our conception of a human spirit and its role in determining human nature and human evolution?
- Do genetic "disease" and "abnormality" have a role in the evolution of humans as beings endowed with bodies and souls?

Eventually teachers can turn to a rather different consideration of model-building by presenting the language of scientists who describe DNA in metaphorical terms – an extreme form of "modelling DNA" from within the realms of poetry and religion. Relevant readings include "Genetics, God, and Sacred DNA" by Dorothy Nelkin (1996) and the metaphorical style used by Walter Gilbert to refer to DNA as the "Holy Grail" (Kevles and Hood, 1992). This mixing of science and the humanities can be used to lead into two additional images as the subjects for discussion: the end-on, rosette-like, computer-generated image of the DNA molecule (Weinberg, 1985) and a photograph of one of the rose windows from Chartres Cathedral (Mâle, 1983). By this time the class should be well-prepared to discuss: how these images serve as simplified descriptions for something that is unseen; how the images are similar and different as models; what the pictures might represent; how the images might be changed to give better conceptions of what they represent.

References

BSCS and AMA (Biological Sciences Curriculum Study and American Medical Association) (1992) *Mapping and Sequencing the Human Genome: Science, Ethics, and Public Policy*, BSCS and AMA, Colorado Springs, Colorado.

Crick, F. (1966) *Of Molecules and Men*, University of Washington Press, Seattle, Washington.

Hamilton, E. and Cairns, H. (eds.) (1961) *The Collected Dialogues of Plato: Including the Letters*, Princeton University Press, Princeton, New Jersey.

Holdrege, C. (1996) *Genetics and the Manipulation of Life: The Forgotten Factor of Context*, Lindisfarne Press, Hudson, New York.

Kevles, D.J. and Hood, L. (eds.) (1992) *The Code of Codes: Scientific and Social Issues in the Human Genome Project*, Harvard University Press, Cambridge, Massachusetts.

Mâle, E. (Wilson, S., trans.) (1983) *Chartres*, Harper & Row Publishers, New York, New York.

Nelkin, D. (1996) Genetics, God, and sacred DNA, *Society* 33, pp. 22-25.
Watson, J. (1968) *The Double Helix*, Atheneum, New York.
Weinberg, R. (1985) The molecules of life, *Science* 253(4), pp. 48-57.

Judgement forming in education

FRANS OLOFSEN

In this time of individualization and secularization, the classical ethics and values in many cases don't give an optimal basis to cope with technical developments in the biomedical sciences, such as genetic modification. It is therefore important that students (from secondary schools and universities) learn to form their own, individually built-up 'judgements' about these developments. Thus, there is a need for a 'pedagogy of judgement forming'.

A recent (October 1995) example of a far-reaching manipulation, which can be seen as representative for many other 'manipulating' techniques and asks for an Individually built-up judgement, is the so-called 'ear-mouse'. This is a mouse with a 'knocked out' immune system, that carries under the skin on its back a plastic mould impregnated with human collagen cells. The techniques used in this case are not new and thus not very shocking, but the image of the mouse with a human ear under its skin are experienced by many students as deviating.
In confrontation with the image of the ear-mouse they react quite the same. Most of them are shocked and display a sort of anxiety. Sometimes there is anger.
Sometimes they just laugh. But when the motives for this sort of research are discussed, their anxiety becomes less and in most cases fades away. With the cognitive thinking, the rational becomes dominant. The possibility of giving humans with or without a damaged ear this sort of 'organic' tissue diminishes the first reaction of anxiety. But the first time they experience inwardly the difference between the 'given' nature (the intact mouse) and the 'made' or 'fallen' nature (the ear-mouse) is a very crucial one. In a sense one can call the inner experience a transcendental, spiritual experience. They feel the difference, for the main part subconscious, between the 'whole' of the (intact) mouse and the ear-mouse.

A mature judgement forming involves both dimensions, which can also be defined as 'vertical' (the feeling of the whole mouse) and 'horizontal' (the feeling of the 'made' mouse). The horizontal process is developed via practical arguments ('with these techniques you can cure people with damaged or lacking tissues') and the 'vertical' process by a sort of intuition of the 'whole'. At the crossover point of the vertical and horizontal dimensions, there is a sort of vacuum. People generally don't know how they can harmonize their 'vertical' intuitions with their 'horizontal' arguments. The process of filling that vacuum is the essence of a modern way of judgement forming. Because in many cases there is an emphasis on the 'horizontal' arguments, it is necessary for a mature judgement forming to accept the 'vertical' dimension fully.

In education there must be more attention given to both dimensions and not only to the one. In this way it will be possible to keep students from following just the 'intellectual' thoughts and models which others have made for them – for example the content of biology books which are sometimes very one-sided. In this context the Dutch *If*gene working group Gene Manipulation and Judgement Forming (Werkgroep Genenmanipulatie en Oordeelsvorming) is developing a school project on gene technology and judgement forming to use in for instance biology lessons.

16 The biotechnology dialogue in the Netherlands

WORKSHOP

HUIB VRIEND
Agronomist
Consumer and Biotechnology Foundation
PO Box 1000, NL-2500 BA Den Haag

Abstract
Recombinant DNA technology and its applications can be judged from different points of view. We can look at the way this technology interferes with nature or the Creation (the moral point of view), at its impact on several economic interests and social structures (the socio-economic point of view) or at the way decisions are made about genetic engineering (the political point of view). These different points of view have their own value and should play a role in regulation, consumer information and, last but not least, commercial and scientific recombinant DNA Research & Development. Using the Dutch dialogue between NGO's and industry and trade as an example, this contribution is mainly focusing on the question how decisions about market introduction and biotechnological Research & Development could be made in a more democratic way.
In my contribution I describe briefly the history of this dialogue. I have used the main items that have been discussed to show how the dialogue works and what the conditions for a successful dialogue are. I end with some remarks about my personal experience.

History: towards polarisation

Soon after the first recombinant DNA experiments in the late 1960's a debate about the consequences of this technology started. It was a technical debate about the potential dangers of these experiments in which mainly biomolecular scientists were involved. At a conference in Asilomar an important group of scientists even decided to put a (temporary) ban on recombinant DNA experiments until the mechanisms and risks were better understood and managed.
However, this ban was lifted almost as soon as it had been put, and in the late 1970's in

the Netherlands a special recombinant DNA committee discussed several issues related to biotechnology. Here again, the debate was dominated by scientists.

It took until about 1985, when the first applications of recombinant DNA technology in medicine and food production became clearly visible, for non-governmental organisations (NGO's) to become involved in the biotechnology-debate. Applications like the recombinant Bovine Growth Hormone (rBST), herbicide resistant crops and a wide range of enzymes in the food industry didn't only raise questions about the safety for humans, animals and the environment, but also raised ethical and socio-economic questions: How far should we go in changing the genes of living beings? Who will be the benefitors of this technology? What does it mean to the producers of several raw biomaterials?

The first reactions of NGO's were quite negative. It looked as if the biotechnological development was mainly dictated by economic interests, and the only benefitors would be the large, transnational agrofood industries. Genetic engineering would bring little or no benefits to consumers and farmers, would be detrimental for small scale farmers in developing countries and animals. Moreover, there was still a lot of uncertainty about the risks for humans and the environment. A rather small, but influential group of people started to organise NGO-meetings, campaigns against unwanted developments and information campaigns for the large public. This group also participated in debates with representants from industry and biomolecular scientists, but soon discovered that those parties ignored the ethical and socio-economic issues and tried to separate the debate about these issues from the safety-debate. Convinced as they were that the technology was relatively safe and beneficial, the industry and scientists saw no reason to deal with gene technology in a different way than with any other production method. As for the NGO's, both the socio-economic and ethical issues and the safety aspects were part of wider, ideological views on sustainable food production, respect for nature and a balanced, more equal economic development.

In the mean time government authorities had started to develop regulation. In 1990 the European Union introduced specific environmental directives for the 'contained use' and 'deliberate release' of genetically engineered organisms. These directives were implemented on the national level in the different member-states. Also in this stage the European Union started to prepare regulations for genetically engineered foods and patents for genes, gene technology and genetically engineered organisms. In this process the interest of the industry had a central place. The right of consumers on information through labelling was completely denied.

Moreover, the commercial developments in genetic engineering was (and still is) dominated by agrochemical and seed companies, which were used to deal with their customers and their interests and had no direct link with consumers. The situation created a lot of frustration amongst NGO's and led to polarisation of the debate.

Knowledge

Amongst the NGO's that worried about the developments in genetic engineering were the consumer-organisations Consumentenbond (660,000 members) and Konsumenten Kontakt (40,000 members). Normally, these organisations base their judgement about new products on independent research and sufficient knowledge about the use of the products. An independent judgement is also important in the lobby-activities for the consumer interests.

Confronted with the gene technology and its applications these organisations concluded that they lacked sufficient information and knowledge to make a well balanced judgement.

As in most European countries, the Dutch government was concentrating its regulation on safety aspects. Other issues were to be solved by the parties involved. On the other hand, the Dutch ministry of Agriculture acknowledged the importance of a more equal position between the industry and scientists on the one hand and NGO's on the other hand. Lack of knowledge on the side of NGO's could easily lead to an unconditioned 'No to genetic engineering' from NGO's because of this lack of knowledge. Therefore, in 1991 the ministry of Agriculture decided to support the initiative of Consumentenbond and Konsumenten Kontakt to raise a kind of 'scientific bureau for biotechnology and food production', called the Consumer and Biotechnology Foundation (C&B).

During the five years of its existence C&B contributed to the development of consumer criteria: safety, benefits, moral criteria, consumer information and choice. C&B was involved in the development and implementation of regulation regarding novel foods (including genetically engineered products) on the national and EU-level. Apart from these activities, C&B tries to improve the transparency around the market introduction of genetically engineered foods and to achieve a more balanced debate in which both moral and scientific arguments do count.

Dialogue

With several applications of genetic engineering becoming ready for the market the nature of the debates changed. The marketers from the industries became involved, and the attention to consumer acceptance increased. Public opinion surveys showed the importance of environmental and consumer organisations as opinion-makers. Probably for this reason the Danish enzyme producer Novo Nordisk organised meetings with NGO's from several EU member-states in which views on environmental risks were exchanged. During those meetings Novo Nordisk did not only try to convince the NGO's about the safety of the recombinant DNA technology they used, but also listened carefully to the arguments of the NGO's. Novo Nordisk understood that a serious dialogue with criticizers could be more fruitful than ignoring them.

In 1991 Unilever had plans to introduce a 'lipase' (an enzyme which modifies fat) in detergents, which was produced with a genetically engineered micro-organism. Unilever wanted to discuss this introduction with Consumentenbond and the Consumer and Biotechnology Foundation. This resulted in a couple of meetings in which was concluded that it would be worthwhile to discuss genetic engineering in a broader context and to invite other relevant food, agro- and seed industries, the retailers organisation and environmental organisations for a consultation group.

Informal consultation group on biotechnology
By giving this consultation group an informal character the participants were enabled to give their opinions freely, without directly facing consequences. This created the possibility to look where the opinions differed and where not. It also created an atmosphere in which it was possible to listen carefully to arguments of the other parties, and to discuss these arguments. This was leading to a better mutual understanding, which is one of the main conditions for finding solutions all parties can agree with.
One of the first and most pregnant issues on the agenda was the labelling of genetically engineered foods. NGO's had been requesting the government authorities for compulsory labelling of all foodstuffs made with genetic engineering. The government authorities replied this would be impossible because a national regulation would have to be notified to the European Commission, and the Commission would certainly reject compulsory labelling. It was estimated that there still would be a long way to go in Brussels before an agreement about a novel food Regulation was reached (which appeared to be right). After long deliberations all parties in the Informal Consultation Group on Biotechnology (ICGB) agreed that easily accessible consumer information about genetically engineered foods is very important for the freedom of choice and the trust of consumers. It was also agreed that in a number of occasions this could be done best by (voluntary) labelling. A common declaration was published which showed in what occasions there was agreement about labelling.

Just before Monsanto notified its application for the marketing of a Roundup (herbicide) resistant soybean under the Dutch Novel Food Act, the ICGB started a debate about herbicide resistant crops. Soon it was clear that the main issue was the contribution of applications of genetic engineering to a (more) sustainable agriculture. As government regulation is focused on the safety of experiments with and products of genetic engineering, there is no existing formal framework for the assessment of the contribution of new products and technologies to (more) sustainable production methods.
In the labelling debate it was possible to convince the representants from industry and trade that labelling can provide for the necessary consumer information and consumer trust, because they are crucial for the marketing of genetically engineered products. That's still talking commercial language. The debate about genetic engineering and sustainable agriculture goes a big step further, because the central issue has to do with

different views on nature or ideologies, which can only vaguely be translated in the same kind of commercial language. Nevertheless, the ICGB is probably one of the few fora where issues like this can be discussed with the different stake-holders, so it's worthwhile trying.

Conditions

I already mentioned the informal character – the freedom to express views without direct commitment – as an important condition for a successful dialogue between different parties about recombinant DNA technology. Of course, such freedom to express views is worthless if the participants are not prepared or unable to be open, if there is no willingness to share information. This is a rather difficult point, because companies will not share strategic and technological information with competitors, nor with NGO's. Also NGO's can have their reasons for not sharing all information, for instance in preparing a specific campaign or article. Therefore, it is important to be mutually aware that the information flow has certain limits and an agreement is made how to deal with information which is shared and marked 'confidential'.
Probably more important than sharing information is the creation of mutual trust. This is more a process than something that can be agreed about. Time is needed to learn about the motives of the participants. Hidden agendas, such as trying to disable NGO's to take a strong opposition against certain applications of gene technology or abusing information for surprising NGO-campaigns against genetically engineered products, are detrimentous for this process. There should also be coherence between the views expressed inside and outside the dialogue-meetings. Part of the process is to develop personal respect for different points of view and moral values and the willingness to discuss these values and the consequences.

Finally, we have to be aware that the dialogue is a means, and not an aim. Sooner or later, the companies and NGO's which are represented will ask the participants to show some results. Without results no further steps. So, there has to be a point in the debate where the informal character ends, and the participants need a mandate to make compromises. The companies and NGO's will have to show a commitment to those compromises. Ultimately, this means the dialogue is not a good strategy for NGO's which have a policy of opposing genetic engineering for moral or ideological reasons.

Personal remarks: try and seek

I am convinced that communication between stake-holders is essential for real progress, in a moral, a socio-economic and a political sense. In a good dialogue arguments and ideas are exchanged, taken serious and judged on their correctness and their value. So far some easy theory. Everyday practice is often more difficult. My participation in the dialogue does not mean I always have complete faith in the outcome. I've got some

personal values about (bio-)technology, food production, social relationships, democratic control and respect for nature, which are my personal drive for doing my work. Maybe part of these values will finally be reflected in the results of the dialogue, and I sometimes ask myself if that part will be big enough. Even worse is when those values get completely faded during the long process of dialogue.
Nevertheless, I think there is not a real alternative. Being realistic one should recognise the potential of genetic engineering for food and medicine production. Then it's worth to try and seek constructive solutions and to realise more democratic control over the technological development.

Note

Huib de Vriend is general policy officer of the Consumer and Biotechnology Foundation and has been involved in the Dutch biotechnology debate since 1985.

17 The genetification of our culture

WORKSHOP

FLORIANNE KOECHLIN
Biologist
Blauenstrasse 15, CH-4142 Münchenstein

Gene technology as a cultural phenomenon: turtles through and through

A famous scientist once gave a lecture on astronomy. He described how the earth goes round the sun and the sun round the centre of the Milky Way. At the end of his lecture a little old lady stood up and said, "What you've just told us is nonsense. In reality the world is flat. It's a plate on the back of a giant turtle." The scientist chuckled condescendingly and asked, "And what is the turtle standing on, madam?" "You're pretty clever, young man, pretty clever," said the old lady, "but it's turtles through and through."

The realities behind the realities are often just those that we have to question. For instance, gene technology has for a long time been not only an activity confined to laboratories and research institutes but also, as impressively documented by Susan Lindee and Dorothy Nelkin (1995), a cultural phenomenon. I should like to pick up two aspects in this article. The first is the way in which images manage to shape our perception. The other is how an activity like gene technology can become a very dangerous justification of eugenic and utilitarian tendencies in society. Both aspects are interconnected.

The photograph of the fourteen-eyed fly

Last year a photograph of a fly with fourteen eyes caused a great sensation. Walter Gehring's research group at Basel Biocentre had discovered the *Drosophila* "master control gene" which triggers eye development. The researcher can produce flies with up to fourteen eyes. Even the corresponding control gene from mouse causes flies to produce extra eyes. Eye "master control genes" have also been found in man. According to the research team this discovery necessitated a rewriting of evolutionary theory.
Other realities are to be found behind the reality of this photograph.

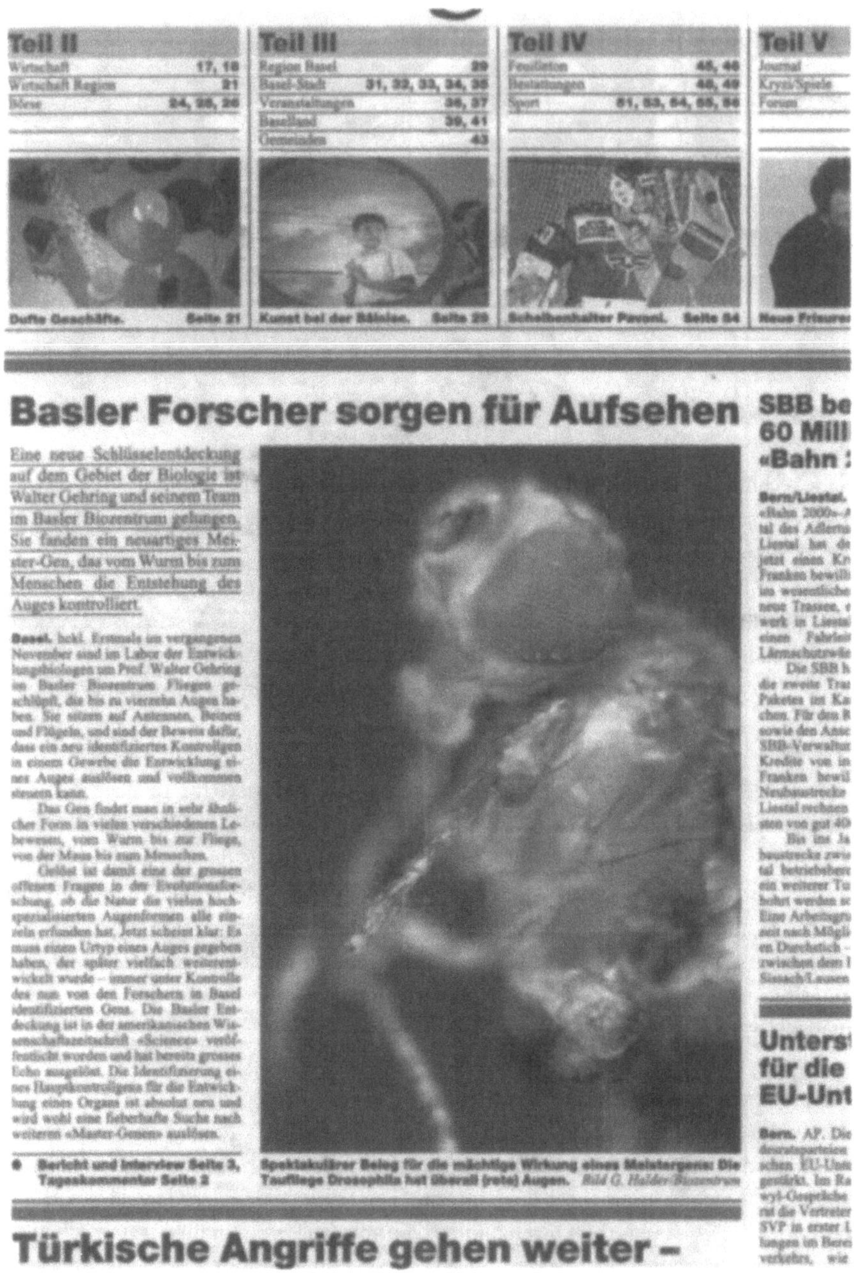

Teil II		Teil III		Teil IV		Teil V
Wirtschaft	17, 18	Region Basel	29	Feuilleton	45, 46	Journal
Wirtschaft Region	21	Basel-Stadt	31, 32, 33, 34, 35	Bestattungen	48, 49	Kryzi/Spiele
Börse	24, 25, 26	Veranstaltungen	36, 37	Sport	51, 53, 54, 55, 56	Forum
		Baselland	39, 41			
		Gemeinden	43			

Dufte Geschäfte. Seite 21

Scheibenhalter Pavoni. Seite 54

Basler Forscher sorgen für Aufsehen

Eine neue Schlüsselentdeckung auf dem Gebiet der Biologie ist Walter Gehring und seinem Team im Basler Biozentrum gelungen. Sie fanden ein neuartiges Meister-Gen, das vom Wurm bis zum Menschen die Entstehung des Auges kontrolliert.

Basel. hckl. Erstmals im vergangenen November sind im Labor der Entwicklungsbiologen um Prof. Walter Gehring im Basler Biozentrum Fliegen geschlüpft, die bis zu vierzehn Augen haben. Sie sitzen auf Antennen, Beinen und Flügeln, und sind der Beweis dafür, dass ein neu identifiziertes Kontrollgen in einem Gewebe die Entwicklung eines Auges auslösen und vollkommen steuern kann.

Das Gen findet man in sehr ähnlicher Form in vielen verschiedenen Lebewesen, vom Wurm bis zur Fliege, von der Maus bis zum Menschen.

Gelöst ist damit eine der grossen offenen Fragen in der Evolutionsforschung, ob die Natur die vielen hochspezialisierten Augenformen alle einzeln erfunden hat. Jetzt scheint klar: Es muss einen Urtyp eines Auges gegeben haben, der später vielfach weiterentwickelt wurde – immer unter Kontrolle des nun von den Forschern in Basel identifizierten Gens. Die Basler Entdeckung ist in der amerikanischen Wissenschaftszeitschrift «Science» veröffentlicht worden und hat bereits grosses Echo ausgelöst. Die Identifizierung eines Hauptkontrollgens für die Entwicklung eines Organs ist absolut neu und wird wohl eine fieberhafte Suche nach weiteren «Master-Genen» auslösen.

● Bericht und Interview Seite 3, Tageskommentar Seite 2

Spektakulärer Beleg für die mächtige Wirkung eines Meistergens: Die Taufliege Drosophila hat überall (rote) Augen. *Bild G. Halder/Biozentrum*

SBB be… 60 Mill… «Bahn …

Unters… für die EU-Un…

Türkische Angriffe gehen weiter –

Figure 1.

Photograph on the first page of the *Basler Zeitung* of 25-3-1995. Drosophila fly with fourteen eyes, as a result of gene technology.

When we look at the photograph of the fourteen eyed fly it gives the impression that we have things under our control. The photograph trumpets the triumph of modern genetics and says more than any words can that geneticists have become "Creators", that they have begun to understand the fly's "switch-gear" and can now make new sorts of flies. For me the photograph had something threatening about it because it showed graphically how we can use gene technology to interfere directly with genetic characteristics of living organisms, be they flies, mice or human beings. Elizabeth Beck-Gernsheim, a German professor of sociology once remarked, "Life still surrounded by the remnants of an almost religious inviolability, at least until now, becomes as technically manipulable as any synthetic material." And the English author John Berger commented, "We see what we look for, and looking is a sort of choice." A photograph entails a choice; it conveys a particular outlook on the world. In this case the photograph is a detailed close-up of part of a fly. Looking at the fourteen-eyed fly I perceive that a fly is made from genes which determine its entire development. Anatomical parts, cells and genes are disposable, interchangeable apparently – and can be commercialised. I remember well another fly photo. In school we had O. Schmeil's biology textbook "Leitfaden der Tierkunde" (An introduction to animal biology). Its coloured pictures made a big impression on me. They included bluebottles, greenbottles, hover flies, dung flies and horse flies; and they buzzed around, landing on flowers, bones, dead birds or simply disappeared into the distance – a wonderful picture. And with the help of little numbers I could find the right descriptions of the flies. Now I wouldn't want to glorify the past, but this picture could *also* have symbolised the

progress of science, a holistic science in which the various approaches interpenetrate and enhance each other. It would be a science that made an effort to understand living organisms in relation to their environment (as shown in the picture) and which in investigating evolution and behaviour took factors in the surroundings into consideration. Here I quote the theologian Günther Altner: "People who understand and treat organisms only as 'boxes of genes' have decided on a view of life in which everything can be reconstructed just as the whims of power wish. But people who see life in its phenotypic uniqueness and wholeness, in its ecosystemic interconnectedness, in its cultural and social dimension (including man-nature interrelations) and as Creation, will demonstrate the advantage of quite other standards for dealing with life and even of a certain restraint towards gene technology." Perhaps the fly photograph will awaken desires in the onlookers. What if we could make not just flies with fourteen eyes but long-lived flies, or not just long-lived flies but long-lived mice and people?
Who knows?

Our genes – our lot?

It seems to me that one reason genetic engineering has such a high profile is because it is connected with mankind's eternal dreams of health, beauty and longevity. Who doesn't want them, who doesn't yearn for a better life? The fascination with gene technology

Figure 3.
This advertising shows a Prometheus-like man reaching for the sky, and entangled in his own DNA-string. It strikes me as an extraordinary symbol of genetic 'can do, must do' fantasy. It also gives a notion that mankind is encaged in his DNA – the genes being the basis of destiny and personality.

certainly has something to do with the little progress which has been made with cancer and other diseases. And now a *Deus ex machina* is emerging in the form of genetic engineering. At least we have our hands on the key. Indeed it is also connected with the fact that the technology promises simple solutions in this terrible complicated world. And it does this without asking us to change our lifestyles, for instance by eating less or taking more exercise. "Defective gene switched off or exchanged – problem solved," or something like it ran the ill-concealed message of the announcement of the above breakthrough.

Research in molecular biology has without doubt made spectacular progress in the investigation of the biochemistry of cell structure. It has also achieved a great deal in the understanding of a particular kind of genetic disease, namely those caused by a defect in a single gene (e.g. Haemophilia B, cystic fibrosis or Duchenne's muscular dystrophy). But together the monogenetic disorders comprise only two percent of all illnesses. The remaining ninety-eight percent of illnesses have far more complex causes. Molecular biology itself increasingly demonstrates that apart from a few rare cases there are no simple "gene switch" solutions like the fly photo suggests, and almost all illnesses and above all behavioural characteristics are based on a highly complex, impenetrable interplay of various genetic and environmental influences.

Politically correct eugenics?

Yet what is stuck in our heads is a picture of man determined by his genes. Our genes – our lot? Apparently biology has become destiny once more and in the old debate over genes versus environment the pendulum has again swung far out on the gene side. New educational methods and equal opportunity in schools? Meaningless if stuttering, learning disabilities and intelligence are genetically determined. Rehabilitation measures for offenders are doomed to failure if aggression and "inability to control impulsive behaviour" are inborn.

On the other hand prenatal diagnostics now investigates the embryonic genome for possible hereditary disorders. At the moment it is still up to the discretion of each individual woman whether to take advantage of prenatal diagnosis. But the pressure on parents not to bring handicapped children into the world will grow, especially considering the huge financial burden on the health system. A mixture of compelling facts ("Something like that would not have occurred with today's medical facilities") and discrimination against the disabled makes it very difficult for pregnant women to opt out of prenatal diagnosis or choose to have a handicapped child. Insurance companies will likewise exert more pressure on parents; in the USA health schemes have already refused to take on children with gene defects. But where is the boundary between "normal" and "abnormal", and, above all, who decides which life has "worth" and which is "worthless"? Should an embryo whose genes show cystic fibrosis or trisomy-21 be

aborted? Are embryos with inborn restricted growth or a defective appetite gene an intolerable burden on the parents? And what about supposed genes for aggression or homosexuality? It is not at all clear whether these genes really exist. But this does not really matter: what man defines as real is real in its consequences. If there is sufficient scientific consensus about the existence of such genes they will be used as basis for selection. Here a new eugenics is beginning to manifest and it is creeping unannounced into the domain of the individual. According to the German sociologist Ulrich Beck it is an "abstract eugenics". "Abstract" ova whose subsequent personality is not yet apparent can be picked out. They don't scream and they cannot defend themselves. In this lies the greatest difference from the eugenics programmes of the Third Reich, and it allows for a far more extensive use of technology.
The Zurich psychotherapist Aiha Zemp knows only too well from her own experience how radically the climate for people with disabilities has changed because of the possibilities offered by genetic testing. She was born with stumps in place of arms and legs. She is used to people meeting her with sympathy or not taking her seriously. But what is new and what has recently begun to happen to her almost every week is hearing comments at the station or restaurant like "People like her would never be born these days." She feels all the more driven to justifying and explaining that she is glad to be alive. It is quite an unbearable situation. Furthermore, aggression towards her has increased. Handicap, says the German sociologist Elizabeth Beck-Gernsheim is no longer destiny but has become something which is not only avoidable but also should be avoided.

It's not only medicine that changes with gene technology. It also dramatically changes how we picture ourselves, nature and life. It affects our concern for our fellow human beings, our responsibility and our freedom. Gene technology is becoming not only a cultural, but also a social phenomenon that would be too dangerous to overlook.
It's turtles through and through.

Reference and note

Nelkin, D. and Lindee, M.S. (1995) *The DNA Mystique,* Freeman and Company, New York.

Florianne Koechlin, biologist with a specialisation in gene technology, responsible for the European coordination of "No patents on life!", Author of *Schön, gesund & ewiger Leben. Bilder und Geschichten zur perfekten neuen Welt der Gentechnologie* (1994, Switzerland)

18 in context – genes, organisms and evolution illustrated through algae and buttercups

WORKSHOP

BRIAN GOODWIN
Biologist
Department of Biology, The Open University
Walton Hall, Milton Keynes, Bucks MK7 6AA, UK

MARGARET COLQUHOUN
Biologist
Kirk Bridge Cottage
Humbie, E. Lothian, EH36 5PA, UK

Abstract
An Algal Order, the Dascycladales and a Family of higher plants, the Ranunculaceae, are used as examples for discussion on the robust nature and intrinsic wholeness of the order of organisms in time and space. The developmental cycle of species, both algae and buttercups, research on morphogenesis, a resultant mathematical model and hands-on experience of plant phenomena all provide discussion material on our experience of the origin of form and the action of genes in the context of organisms, species and wider taxonomic groupings.

Introduction

This Workshop focused on the question: How are we to understand gene action within the context of developing and evolving organisms? This is clearly relevant to an understanding of the consequences of transferring genes between species, often distantly related, as is practised in biotechnology. Our primary concern was to broaden conceptual horizons through introduced examples and discussion, leading beyond the limitations of genetic reductionism with the aim of recovering an understanding of organisms as transforming wholes, rather than addressing detailed practical questions.

The workshop participants were around 15 in number and we began with introductions of who we are and what we do. Brian Goodwin introduced himself more extensively, with details of his biography, as a biologist who had grown up and studied in Canada with an interest in evolution and developmental biology. He later studied mathematics in Oxford and subsequently worked for his PhD with Prof Waddington in the Institute of Genetics at Edinburgh University in the 1950's. This was a tremendously exciting time; The double helix of DNA had just been decoded and Jacob and Monod described their regulating systems. Inspired by Waddington, (who had moved from Geology to Biology), and working as a theoretical biologist, Brian's main interest, then as now, was in the dynamics of the developmental process, and in the dynamics of the regulation of gene activity. Fairly early on in his career it became clear to him that the conventional DNA story and Darwinism does not describe all the phenomena. This led to a life-long exploration of alternatives, some of which research was shared in this workshop.

Our concern in the workshop was to look at the context of the genes: a) in the context of the organism (developmental biology) and b) in the context of evolution (evolutionary biology) using an Algal Order, and the Buttercup Family to illustrate ideas and stimulate discussion.

The Dascycladales – stability of form over 600 M. years

The algal order Dascycladales provides an excellent and relatively simple example to explore some of our questions and was used here to examine the relationships between gene action, morphogenesis, and evolution. The Dascycladales are giant unicellular green marine algae whose fossil remains can be traced back as far as the Cambrian, some 600 million years ago. Calcification of the cell walls results in detailed casts of morphology, which reveals that most of the surviving species have structures very similar to fossils between 350 and 600 million years old. We looked at slides of a large variety of both fossil and living forms and were invited to see what they have in common. It was not difficult to find that they all share a common basic bodyplan: a central stalk with whorls of laterals. This common pattern holds true now for many species in space as it has held true to type in time since the Cambrian – even in the most recently evolved species which shed the laterals soon after they are formed, making no use of them in the adult. A question thus arises: Why is this basic form so stable? Is it due to natural selection? Is it because of a specific genetic program of morphogenesis that is conserved? Or is it that the basic organisation of this class of organisms is such that this form is 'generic'; one that arises more or less inevitably from the dynamics of morphogenesis in giant unicellular algae?

Origins of form and "fields"

Before we could go more deeply into these questions a discussion arose on the origin of form in organisms – whether environmentally or genetically determined and concerning that which holds things true to type. With Darwinistic assumptions any form is possible in evolution. An alternative hypothesis would be that there are limited sets of forms possible for the expression of each group (or type) of organisms, the group as well as the individual organism having its own intrinsic value. This idea was compared with the limited set of possible states for, say cloud morphology, (after Goethe) and led to further discussion on the concept of "field". A field was defined as being "a domain of relational order that changes in space and time according to dynamic principles". This could hold true for all levels of organisation from clouds to crystals to plants and from individuals to populations to higher orders of organisms. It is an idea that has been brought from physics into biology.

In considering biological fields within single organisms, we need to understand the way the cells are organised, especially in such unicellular organisms as the Dascycladales. We can ask the question; does the Darwinistic hypothesis of "any shape" fit the phenomena or does the evidence suggest organisation of a particular type or constraint in the patterns they express? Here we are moving into a pre-Darwinian tradition of rational morphology prevalent at the end of the 18th century in people such as Goethe and his contemporaries, in Georges Cuvier, Geoffroy Saint-Hilaire last century and D'Arcy Wentworth Thompson, C.H. Waddington and J. Needham this century. One can even trace this way of thinking back to Aristotle.

We explored in conversation the difference between inner and outer constraints, the idea of intrinsic constraints of organisms and the need to always consider the context of the organism and how biology has tended to be a historical subject based in contingency and narrative. The theory of natural selection is not explanatory, but merely historical.

Acetabularia – life cycle and organism

We then returned to the Dascycladales and were introduced to the life cycle of one of them, Acetabularia acetabulum. Here the term "organism" is synonymous with "life cycle" as the organism is constantly changing in order to be – to express itself. Its form, however, is relatively stable and has something in common with all the Dascycladales. (Fig. 1) If we concentrate on the first part of the life cycle (See Fig) we see isogametes come together and the resultant organism grows taking 6 weeks to reach the cap stage from which the alga gets its common name, (the Mermaid's Cap). If the cap is cut off it regenerates by the same process as occurs at an earlier stage in the life cycle. The tip grows, extends and flattens. Laterals form from buds. Eventually the cap forms and the laterals drop of. The laterals are made of an extension of the cell wall. Now why are all

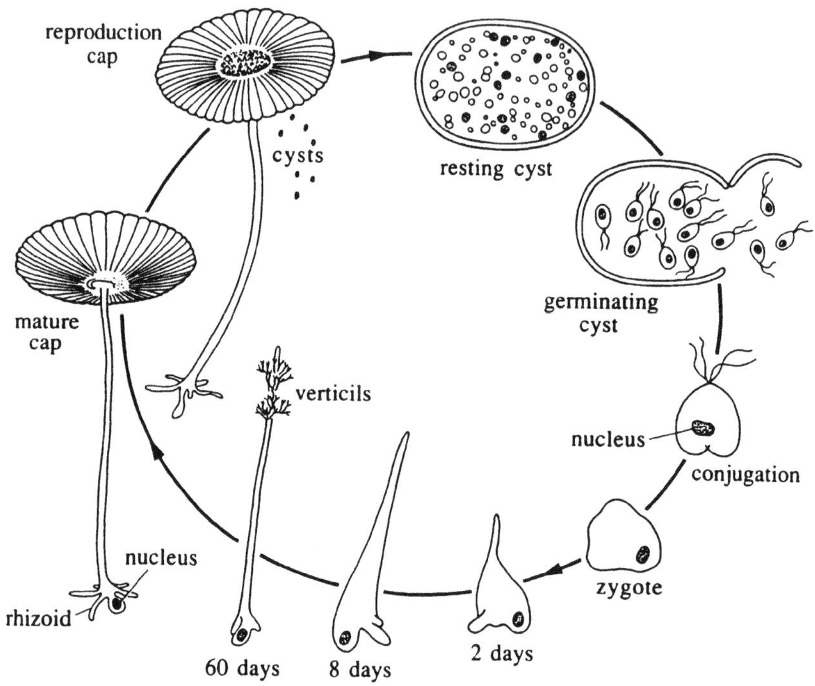

Figure 1. Life cycle of Acetabularia acetabulum

the members of this taxonomic group doing the same thing after 600 million years? Why is the gesture of lateral formation so stable and why in particular in this organism which apparently "does not need them" in order to reproduce as do its ancestors? Is it because they are evolutionary relicts, in which case one would expect them to have degenerated (according to Darwinism) or is it because they are intrinsic to the organism?

Modelling the genesis of form

These questions have been explored in the research of Brian and his colleagues at the Open University. He described some experimental observations on the growth of cell walls, the properties of the cytoplasm and the vacuole, and the role of calcium in the morphogenesis of the algae. (Details of this can be pursued by interested readers in the book "How the Leopard Changed Its Spots" (Goodwin, 1994)). An understanding of these changing phenomena was used as the basis of constructing a mathematical model of morphogenesis in this type of unicellular organism. The model was based on the interactions of calcium with the components of the cytoskeleton (primarily microtubules, microfilaments, enzymes and the calcium regulatory system), the growth of the wall in response to changes of state in the cytoplasm, and the pressure exerted by

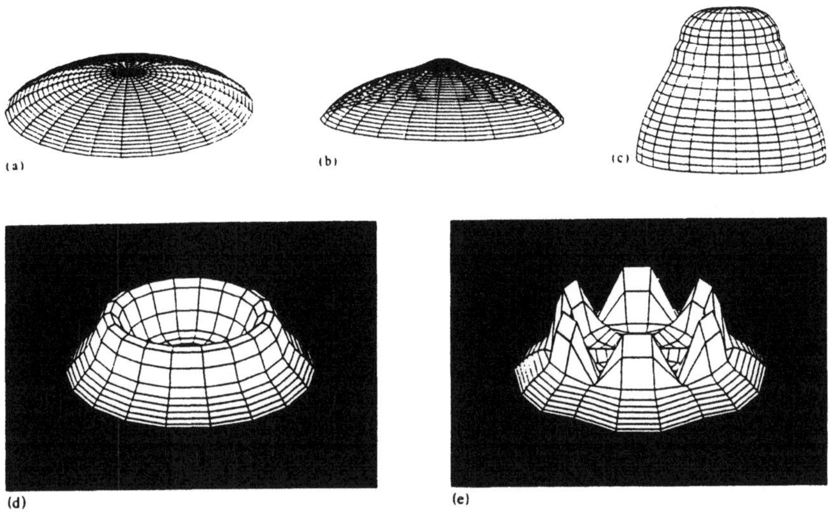

Figure 2. Mathematical model of morphogenesis in a unicellular organism

the vacuole on the cell wall. This model was then simulated on a computer to see what forms it would generate. The structure that developed was a moving picture of the generic form of the Order: starting from either a sphere (the zygote) or from a hemisphere (a regenerating tip), a growing stalk was produced which generated at its tip a sequence of whorls of laterals (Fig. 2). The model provided insights into the reasons for this form in terms of a natural cascade of symmetry-breaking events that spontaneously followed one another. These were: 1. emergence of a growing tip from either a sphere or a hemisphere; 2. flattening of the initially pointed tip because of a change in the pattern of calcium and consequent changes in the state of the cytoplasm, resulting in changes in the shape of the cell wall; 3. a consequent further change in the spatial pattern of calcium that resulted in the production of a whorl of laterals; 4. repeat of the sequence 1 – 3 to give a series of whorls. No change of parameters was necessary in order for this sequence to occur, so no genetic program was required to generate the basic form.

It was, however, necessary for a number of constants describing properties such as the effect of calcium on the cytoskeleton, regulation of calcium in the cytoplasm, elastic properties and behaviour of the cell wall, and so on, to be within a certain range in order that the generic form be generated. These are determined by gene products. Hence the genes play the important role of assuring that the whole system is within a particular domain of what is called "parameter space" so that the basic form common to all members of the algal order be produced. If these gene-determined quantities fall outside this quite large domain, then morphogenesis fails. Thus genes stabilize the conditions for the occurrence of characteristic patterns. However, they do not determine the properties of the integrated system that is the generator of form. This is an emergent property of the cytoplasm that arose during evolution and is inherited via the cytoplasm contained in the gametes. *Thus genes define necessary conditions for particular forms to appear, but they are not the causes of form.*

Genes as networks

Discussion followed on how this has been revealed by many examples showing that the genomic map is not identical to the morphological map. The genes act as a network. There can be a lot of change in the genes but as long as the overall network hangs together, form is conserved. Different genes can produce the same form and the same genes can produce different forms in a different context (organism) e.g. the retinoblastoma gene. How it manifests is determined by the context in which it finds itself. This is why transgenic organisms are very risky. We also realised the human genome project seems to be hiding some truth here as it will never be able to yield the networks – how the genes interact with one another – to generate us!

The conclusions of the research work demonstrated can be summarised as follows:
1. It is not possible to reduce organisms to the actions of genes. There are principles of organisation (e.g., the eucaryotic cytoskeleton and its relation to calcium, cell wall properties, etc.) that need to be understood at a different level from gene activity (which defines molecular composition) if morphogenesis is to be understood.
2. Evolutionary change in organisms of a particular type (genus, family, order) is constrained by these organising principles. Parallel and convergent evolution within such types is common; that is, species with similar morphology to early (fossil) species are likely to be found, as they are in the Dascycladales.
3. Genes provide the molecular materials that are necessary for the organised structures (cell wall, cytoplasm, membranes, etc.) that make up the organism; but these structures are emergent forms that have arisen during evolution and they are inherited from generation to generation. They are not produced *de novo* from their constituents.

A detailed account of the Dascycladales and the model that describes the morphogenetic sequence leading to the generic form can be found in: "How the Leopard Changed Its Spots" (Goodwin, 1994).

The model-making activity of human beings

The workshop participants then went into animated discussion about the relationship of the model to the reality (on which level?) and the role of the genes in the whole process. For human beings to understand the dynamics of a system we have to make models which integrate the phenomena we perceive. Human beings doing science are making models all the time. Mathematics can be used as a descriptive model of excitable media such as the algal living cell in terms of fields using higher order concepts. Here, as in ordinary daily model-making, the concept has to match the percept at the level of behaviour of the organism we are studying. We, as scientists, choose the level of study and define the parameters out of our observation of the phenomena. Mathematical model-building is a way of testing the hypotheses with respect to the sufficiency of the conditions and any "explanation" arising out of the model-building is true according to the sufficient conditions.

Leaving the field of the Dascycladales, but staying with the question of the model-making process of the human being, we jumped rapidly forward in evolutionary time to a remarkably labile and yet recognisably stable taxonomic group of higher plants, the Ranunculaceae. The Buttercup-Family has been recognised and described as a taxonomic group for well over 200 years. How was this done without the modern DNA sequencing techniques? What is it within us as human beings that makes it possible for us to recognise order in the world of living organisms – or in the inorganic world for that matter? How is it that we are able to classify organisms into coherent groups and what has this order we can readily perceive got to do with the genes of the organism on either a developmental or an evolutionary level?

Reference was made to the quotation in the morning talk by Ernst Peter Fischer of Johannes Kepler on his process of gaining knowledge: *"We gain knowledge when we combine our perceptions with mental ideas and assess them to be in agreement. What is inside will then light up in the soul"*. This became something of a theme for the remainder of the workshop.

Recognising intrinsic order in Buttercup leaf sequences

Everyone was given an envelope with photocopies of all the leaves from a single specimen of two different buttercup species all mixed up. They were asked to order the leaves. With a little bit of help offered to one or two participants in a very short time everyone had managed to put the leaves in the order in which they had grown on the plants and according to the appropriate species. The aim of this exercise was to explore the activity of "model-making" within ourselves and to try to build a bridge to the order perceived by a wide variety of taxonomists through the last two centuries who have worked on and found remarkable consistency whatever method they use within the taxonomic group of the Ranunculaceae. It was also our intention to try to become more conscious of the process of how we create order ourselves in our activity of perceiving it within the world. *"The aim of the conference is to mobilise people"* was quoted from the conference leaflet.

The questions were raised: what is a taxonomic group and what are the constraints that hold it together? Is it simply due to the random mutation and natural selection postulated by neo-Darwinism or is there an intrinsic "belonging togetherness" that creates a boundary to the "field" of the Family of Ranunculaceae? How do we know that things belong together? In other words, how do we find the boundaries and constraints within a group of organisms, recognise and work with this on a phenomenological level? What defines the boundary? Is it we ourselves or is it something truly intrinsic to the organisms? Are we reading nature's book when we do taxonomy or are we imposing our own thought order on the world? Aristotle stated in de Partibus "the method then that we must adopt is to attempt to recognise the natural groups – each one of which combines a multitude of differentiae and is not defined by a single one as in dichotomy".

The process of discovery

The second day of the workshop began with an exploration into the cognitive process based on the work we had done together the day before together with a short introduction to Goethe's contribution to science. We recreated together the activity of ordering the leaves correctly of buttercup plants by describing how we had laid out the leaves on the floor in disorder (relative chaos) and how most people had managed to very quickly arrive at the biologically-correct order despite the fact that most of us had never seen the plants or these leaves before. How was this possible? It was agreed that we had used something like an innate or intrinsic intuition to organise them according to their similarities and differences – an intuition that allowed us to perceive a biologically true relationship in the way the leaves had grown on the plants.

It was brought to consciousness by one of the participants that the forms of all the algae we had seen the day before were similar in structure to the forms of higher plants. Similar forms can thus be seen in both algae and higher plants – organisms with a single cell that have been around for 600 M years and rather more recently evolved and very differently organised multicellular higher plants. The question arose of whether there are restrictions to the number and variety of possible patterns that can be developed in plants – at any stage of evolution and regardless of the environment (land or sea) or time (long ago or recently) (Goethe had found something similar with mosses – as if they did in miniature what the rest of the Plant Kingdom did in large). We seemed to be recognising something similar in pattern formation on many different levels of plantorganisation. Where does this recognition or realisation come from?

We went back to the Acetabularia story of the previous day and asked what was the aim of the experiments? What was the role of the scientist as mediator between the phenomena and the model. We explored, together with Brian, the process he had been engaged in on his journey from observation of the algal organism and its growth process including lateral whorl formation to the creation of a mathematical model. When asked how he had done it Brian said: "It's difficult to say. None of the separate phenomena gave rise to the model. It was a process of synthesis which arose partly in consultation with a Toronto physicist. I came up against a major obstacle in the mathematics of the model. I could not get an excitable medium. I went for a swim and during it I had some inspiration. I went back to the laboratory and it worked like magic. Some synthesis had occurred in the meantime." Other examples of scientific discoveries being made in this way were offered by a number of people – particularly the discovery of the structure of the benzene ring by Kekule.

Grasping the essence

Our explorations into the creative process of the scientist were undertaken with the aim of looking at how science is done and with the question of how a scientist comes to the idea of a transgenic organism at all. What sort of process does he or she as a human being have to go through when engaged in the genetic engineering or modification of organisms and how, where and on what level do ethical considerations play a role in this process?

Spurred on by Johannes Wirz' discussion of his personal journey as a scientist in the morning and by a contribution of one of the conference participants in the plenum session, Margaret Colquhoun then shared something of her biographical journey as a biologist. She told how she is one of those people who love to find patterns and order in the chaos of the world and quoted from Blacklith and Reyment's book on Multivariate Morphometrics, which had been a key work for her while studying for her PhD in Edinburgh in the 1970's:

"The desire to abstract from the great variety of living organisms forms which are essentially harmonious, aesthetically superior to other forms, or else capable of abstraction as the archetypal form of some wider category of shapes and sizes, is very deep seated in human behaviour."

and from Dobzhansky in 1975 who said how it is easy to study multiformity in nature but to *"grasp the essence is a greater achievement than to behold it's evanescent expression"*. This was also Goethe's aim.

What does "grasping the essence" really mean? Has, as might be supposed from our understanding of Kepler in the quotation cited above, the very process of knowing anything to do with the order that is inherent in the organisms around us and their evolution? Has the lighting up of the soul that Kepler described anything to do with our matching our field of knowledge or comprehension with the intrinsic order within an organism's field of existence on whatever level we might be trying to study it – and are our successes in describing this or in making some kind of model due to other people also being able to recognise something of a true interpretation of the phenomena which also lights up in their souls?

Is doing science humanity-enhancing?

Margaret then went on to describe her journey from originally working as a Research Associate at Edinburgh University to the research she has been doing in recent years as a Goethean Scientist. She had worked for her PhD on genetic variation in island populations of small mammals off the coast of Scotland and Wales with the question of whether these animals were glacial relicts or had been introduced to the islands by mankind in more recent times. She sampled populations on the mainland near to the islands as well as taking animals from the islands and back in the laboratory, measured their gross morphology, sampled blood and organ proteins using starch gel electrophoresis, measured many sizes and the absence and presence of bumps and holes on the skulls. All this data was fed into a computer program based on multivariate morphometric analysis, which has arisen out of the premise, first defined by Aristotle, that every organism can be definedly described by number. The results showed that the voles from one island were glacial relicts – they also looked like and bred more easily with the Alpine voles – and the others were probably introduced more recently by man. It also showed that the genetic variation on one level of organisation of the organism was different from that on another level, i.e. organisms are not uniformly hetero- or homogeneous, as had been expected at that time, on all levels of genotypic expression (all the characters studied were genetically determined.) This work with small island mammals in the field had been wonderful as were all the ideas behind and around the evolutionary questions were very exciting but the research had involved killing enormous numbers of animals to get statistical reliability. Many of the results that arose could have been extracted from observation of the animals in their environments without going through the analytical and statistical procedures that were involved. She described what the research had done to her as a human being – how, after a while, she just could not kill any more animals, became nauseated by the smell of ether and began to question the necessity and justification of the whole scientific process being unable to separate the distasteful side of that particular scientific process from herself as a human being. Doing the sort of science she had been involved in had been, for her, like going down a dark and narrowing tunnel within which process she became progressively less and less human. (A similar experience had been described by a conference participant in one of the plenum sessions and was echoed by another member of our working group) Margaret was fortunate, then, to have encountered work with very deprived children. Through them she rediscovered her love of nature. She then contrasted this experience of doing science conventionally with the experience of doing science Goethe's way, which she has been doing for the last ten years and which, by contrast, has been experienced as a richly fulfilling and humanity-enhancing activity.

Through the work of Goethe and the subsequent development of his science as a methodology by Rudolf Steiner it has been demonstrated time and again that it is possible to do science without necessarily always destroying the organism which one is endeavouring to understand. Through becoming conscious of the process of doing science while actually doing the research itself, and by asking questions appropriate to the organism, means that the self-development of the scientist becomes intrinsic to the process of doing science altogether. This can lead to a morality or experience of ethics which grows out of the very subject which is being investigated. One starts to be able to own the whole process and to take responsibility for what and how the scientific results are revealed. One begins to acknowledge, like Kepler, the role of the first person and their satisfaction in the very doing of the research from the phenomenology of the facts through the selection of what to investigate to the actual deed of making a model, whether inner or outer.

One is no longer divorced or separated from Nature – "doing a job, the responsibility for the outcome of which is not mine". If this way of working could be taken into the realm of molecular biology and genetic manipulation then perhaps the ethical questions could be dealt with on a real level of human cooperation with, and consideration of, the rights of the natural world.

Archetypal plant and archetypal organ

Going back to Goethe and the buttercup leaves something of Goethe's experiences on his Italian journey were described – how he had seen plants high in the mountains that were similar in form and yet very different from those he knew to be the same species as those growing where he lived in Germany. He slowly realised that there is no such thing as an "original plant" from which all the others derived but rather that all plants have something of an ineffable/intangible "something" in common which allows us to recognise that they are plants. He called this spiritual idea "the Archetypal Plant" ("Urpflanze"). Likewise, when one goes more deeply into the transformation of the leaves on the stem of a flowering plant such as we had seen the day before, (Fig 3. Ranunculus sarduous, a species of yellow buttercup), we find there is also an intangible something, the "archetypal leaf" that we can experience (with the harmonious or musical part of ourselves) as being a non-physical, which all the leaves have in common and yet is not entirely expressed by any one of them.

The first person to describe that all plant organs have common origin was an Italian called Cesalpino in the mid-16th century. Linnaeus then described the transformation from leaf to bud to flower within the plant as a "Metamorphosis plantarum". For Goethe, having immersed himself in Linnaeus' works, the metamorphosis was not just a description but a living experience. He once, sitting in front of a rose, closed his eyes and then "saw" all the plant parts transform one into another "with inner necessity and

Figure 3. Ranunculus sarduous, leaf sequence

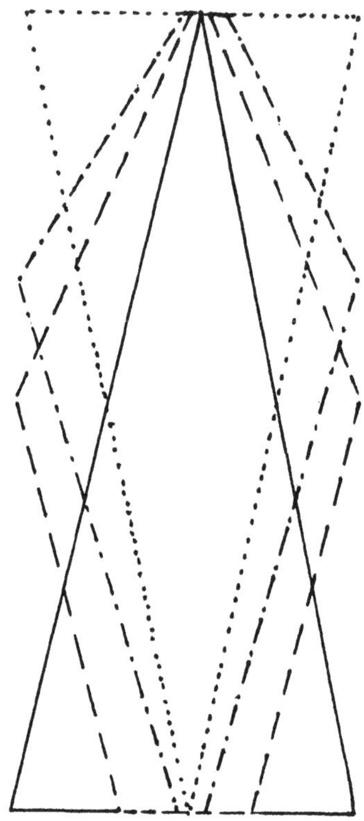

Figure 3a. Ranunculus sarduous

.... pointing
–. – . – . differentiating
– – – – – spreading
——— stemming

truth". The experience of the moving pictures he saw within plants he set out in his poem, "The Metamorphosis of Plants". We discussed how something of this inner experience being externally expressed lies at the basis of the model-making of most scientists whether they are conscious of it or not and whether they express it in poetry or in mathematical models.

Laws in the leaf sequence

Goethe had described the movement of a plant through the leaf sequence towards the flower as a "refining of the juices" in a continuous time stream. In the 1960's Jochen Bockemühl, working at the research laboratory there at the Goetheanum in Dornach, had discovered and described how this stream is a highly differentiated process – a fourfold differentiation in the field of the growing plant. We looked at a number of species of buttercup leaf sequences and found the metamorphic movements they had in common in their leaf transitions were differentiated as follows. The first activity, nearest to the roots was one of stretching the stem, the leaf blade being rather small and simple. He called this activity of expression "Stemming". Moving up the plant (See Fig. 3a) the stem slowly decreases in proportion to the blade as the mass expands into a leafy plane. This activity is called "Spreading" and is polar opposite to the activity of "Differentiation" of the leaf blade higher up the stem when the leaf becomes finely divided and more characteristic of the species. There then follows a contraction as the divided parts are reduced in size and number, the mass decreases, the petiole disappears and finally there remain only tiny pointed leaves sitting tight against and wrapping round the stem of the plant under the flower. This last activity is known as "Pointing". Participants experienced the most harmonious leaves as being those in the middle of the sequence where all four activities seem to be in a kind of balanced tension. It is in this region that we find the most species specific leaves. However, in most plants it is not until the flower is open that it is possible to identify exactly which species a plant is.

Flower parts and gene action

We went on to look at the flower parts by observing how, in a peony or Helleborus foetidus (Fig. 4), as we approach the flower at the top of the stem, the leafy or bladal part of the leaf disappears together with the petiole and the sides of the leaf bases gradually expand in several steps to form the petals. Going further inside the flower we find petals with their edges rolled inwards and revealing the beginning of anther formation, before the stamens themselves are fully developed. It was such "in between" forms that revealed to Goethe the relationships between all the flower parts. All four possible flower organs were described in the workshop with details of their position, usual shape name: sepals, petals, nectaries and stamens with the carpel or beginning of the fruit being in another layer in the middle of the flower.

Figure 4. Leaf sequence from Helleborus foetidus, between leaves and flower

Brian then went on to describe research work on homeotic mutants of Arabidopsis thaliana (by Meyerowitz et al, 1991) in which a model has been created showing how all the "modified leaves" of the flower can be explained by a sophisticated genetic map – at least in Arabidopsis. So again we find that the genes "stabilise", in this case, the form, colour and number of flower parts – that which characterises the nature or specificity of species or Genus. But the way higher dicotyledonous plants, including each member of the buttercup family, goes through its differentiated expression of the formative movements in the leaf sequence is equally characteristic of that species – i.e. is inherited – but in a much more fluid manner. This was observed in the workshop by looking at several different species.

Leaf – flower relationships in the Buttercup family

Margaret then described how, in her research into the buttercup family it had become clear that there is a relationship between the flower parts and the differentiated patterns or style of progression through the leaf sequence in terms of the dominance of formative movement expressed there and those organs which were most predominant within the flower. Plants with a stemming tendency pushing into the leaf blade such as Clematis

Figure 5.
Clematis vitalba, leaf and flowers

Figure 6.
Aconitum napellus, leaf and flower

Figure 7. Adonis aestivalis, leaf sequence (left) and flower (right)

vitalba (Fig. 5) have flowers which are composed of a super abundance of stamens i.e. dominance of the stemming activity within the leaves is accompanied by an exaggerated development of stamens (to take over petal function) in the flower. Likewise, in plants where the pointing gesture is exaggeratedly developed within the leaves, the calyx of the flower turns from green to coloured during development and takes on a petal-like function e.g. Aconitum napellus (Figure 6) The only genus within the family which has "normal" flower part expression also has very "balanced" leaves. Within this Genus, Adonis (Fig. 7) at the heart of the Buttercup Family are the only red European members of the Family. Adonis (vernalis and aestivalis) also happen to be used for heart remedies!

The whole of the buttercup family can be ordered according to these principles. The way the plant grows, their habitat and relation to other plants reflect in gesture the same tendencies found in the leaves and flowers. Photographic examples of the main European genera were shown as slides and their habit described. This taxonomic ordering of the Buttercup Family, which was pursued from a Goetheanistic viewpoint during three years of research at the Carl Gustav Carus-Institut in Oeschelbronn, Germany, turned out to be identical (except for one species) to that described by Linnaeus over 200 years previously. Again it became clear that we, as human beings have the tools to discriminate between form and patterns and to resynthesise our relationships to a very high degree within ourselves if we immerse ourselves in the "field" of the organism or group of organisms. The boundaries to each group, be it an Algal Order, Buttercup Family, Genus or Species, then become as tangible to our perceiving as those boundaries of a single physical plant. They are experienced as edges of the inherent or intrinsic order of a group of organisms among the chaos of all the separate individuals. This is a similar experience for the researcher on the level of a Family to that which we had on the first day when we found the order in the leaves of one plant individual. The beginning of Goethe's poem of the "Metamorphosis of plants" conveys this experience beautifully:

"Confused may you be my beloved,
By the thousandfold mixture of flowers in the garden
You hear many names sounding one after t'other,
With terrible tones in your delicate ear.
All their forms are alike, yet their's none like another,
The whole choir is pointing to a secret law,
To a holy riddle. O if only, dear friend, I could in a word
Convey to you now the happy solution."

Conclusion

In conclusion, conveying the "happy solution" of the order in the world, the secret laws of taxonomy to be imbued into the art of model-making, have a lot to do with our own particular process of appreciation of who or what that group of organisms is expressing in the world – with its intrinsic essence. It is clear that the genes play an enormously important role in maintaining and stabilising that essence in the organism, species, genus or family, but they cannot, we believe. explain the origin of an Algal Order or the differential expression within the wholeness of the Buttercup Family. This cannot be understood in terms of Darwinian evolution and asks us to widen our consciousness to embrace something like an archetypal creative order in the world which appears to be experienced within all the groups or forms we have encountered so far at one level or another, whether they are leaves of a plant, plants in a family or algae in an order, but expressed in an unique and specific manner. We have within us the capacity to perceive the order to our own satisfaction, which itself is an echo of that order in the world, i.e. taxonomy is inherently both within and outwith us and, in Kepler's words matching the two "allows something to light up in our souls".

Given this we would plead for such a methodology of research to be taken up in the fields of Molecular Biology and Genetic Manipulation so that the great wonders of modern technology may be married with a reverence for the intrinsic value of every organism under consideration.

References

Goodwin, B.C. (1994) How The Leopard Changed Its Spots, Weidenfeld and Nicolson, London.

Meyerowitz, E.L., Bowman, J.L., Brockman, L.L., Drews, G.N., Jack, T., Sieburth, L.E., and Weigel, D. (1991) A genetic and molecular model for flower development in Arabidopsisthaliana, *Development Supplement* 1, pp. 234-257.

19 Transgenic plants – consequences and impacts for production and ecology

WORKSHOP

JOS VAN DAMME
Population geneticist
Netherlands Institute of Ecology, CTO
PO Box 40, NL-6666 ZG Heteren

BEAT KELLER
Plant breeder
Eidgenössische Forschungsanstalt für Agrarökologie und Landbau
Reckenholzstrasse 191, CH-8046 Zürich

Reporter: Meinhard Simon (D – Konstanz) and Martin Keller (CH – Zürich)

Abstract
In this workshop plant breeding by genetic engineering, risk assessment and related questions and problems were discussed on the background of a short history of plant breeding in general. The aim of breeding has always been to improve the quality of cultivars. This is also the major aim of genetic engineering which may, however, include risks such as gene transfer of unwanted traits to related wild types or changes in toxicological and allergenic potentials of plants. Risk assessment has to include the probability of escape of the gene introduced and its potential effects on the organism, on the ecosystem after a possible gene transfer and eventually on evolution. It is still very difficult to predict any risk of gene transfer even though some observations suggest that it may be faster than anticipated.

The topic was introduced by Beat Keller in the first session. He stressed that any discussion and arguing on genetic engineering in plant breeding can only be done on the background and in the context of classical breeding in general.
Breeding has a long history starting some 15,000 years ago and it has been always its major aim to get plants adapted to human needs such as nutrition, medical application,

raw material for clothing, construction and energy. Such aims to improve the quality, e.g. for nutrition, included refining the nutritive quality of cereals and to increase the crop of seeds per stalk or the crop yield in general. To improve the resistance of cultivated species against pests and pathogens became an increasingly important goal of breeding. It has to be kept in mind, though, that often such "positive" traits as resistance are linked to "negative" traits such as reduced yield which are difficult to separate out. During the course of this long time period of selection the traits of cultivated species have changed very much compared to wild types. New technologies, not only that of genetic engineering but refined methods to select and optimize for the appropriate varieties, accelerated this process during the recent past considerably. As a consequence, the number of genes per cultivar for which the breeding process selected increased considerably even though in total the gene pool and biodiversity of cultivars was reduced.

Genetic engineering aims to further optimize the breeding process and to put it into a completely rational and manageable context. Hence, breeding with all its methods and approaches has been always a mirror of the social, economical and ecological developments in our society as depicted in the following scheme.

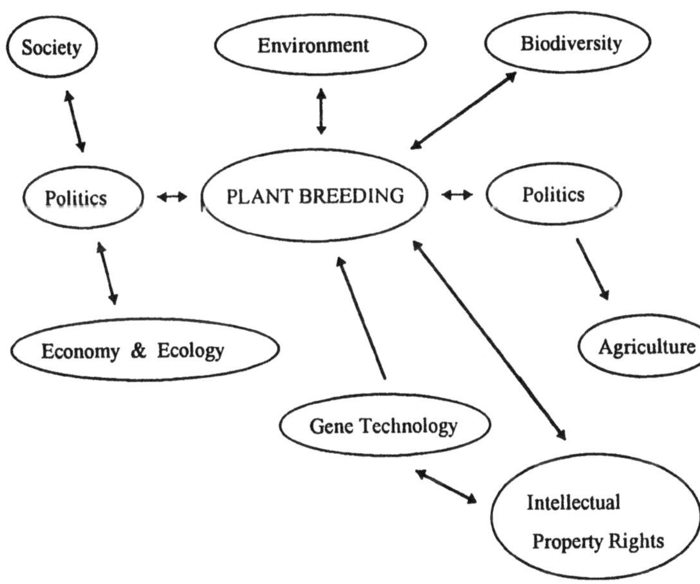

Chances of genetic engineering in plant breeding therefore, including the following points:

- To extend the realm of utilizable genes for traits as resistance, quality and storage losses.

- To reduce the application of pesticides due to inherent resistance of cultivars against pathogens and pests
- The combination of traits is more precise than with classical methods and also includes combinations wich have been impossible otherwise
- The process of breeding might be accelerated

Genetic engineering of plantbreeding, also potentially includes risks:

- Gene transfer of unwanted traits to related wild types (e.g. herbicide tolerance). It has to be kept in mind that genetic engineering does not change the probability of gene transfer but the spectrum of genes transferred including ones which are new in this context.
- Changes in toxicological and allergenic potentials of plants because some proteins involved in resistance reactions are toxic and others exhibit homologies to allergens.

The second session was chaired by Jos van Damme who gave an introduction to risk assessment of transgenic plants.

The risk assessment has to include the probability of escape of the gene and its potential effects on the organism and that of the transgenic organism on the ecosystem and possibly on evolution. The most important result of the assessment is the judgement of the risk, individually, socio-politically and for the environment.
For a careful assessment the probability of escape has to be set to 1, e.g. ultimately the gene will be transferred, either horizontally or vertically. Vertical gene transfer means, that genetic information of one organism is transferred to another by sexual reproduction, e.g. from a crop plant to a wild relative by outcrossing. Horizontal gene transfer means that genetic information is taken up by any non-related organisms, e.g. from plants to microorganisms. It is nearly impossible to predict the time scale of escape which depends on the phylogenetic relationship of a crop species to wild species in a given ecosystem. Vertical gene transfer is highly probable and often yields hybrids that are fertile to some degree. It is difficult to predict exact figures of hybridization but 80% of crops have hybridization events in their breeding history. Effects of horizontal gene transfer for the ecosystem are to be expected more on an evolutionary time scale even though direct effects may occur fairly shortly if the selection pressure is high, e.g. under the continuous application of herbicides. The possibilities for horizontal gene transfer are quite limited, since a vector like a virus or a bacterium is required to transfer genetic material from somatic tissue to somatic tissue. Introgression through sexual reproduction of genes of cultivars into related wild species an vice versa have been observed, like in maize, carrots and lettuce. Even more difficult than the frequency of gene transfer itself is to predict its potential effects, including also long-term consequences on an evolutionary time scale.

The judgement of a potential risk includes many criteria which we get from experiences, experiments, traditions, expectations, socio-political standpoints, etc. Ultimately, however, the judgement comes from a moral attitude and conviction in ourselves which takes into account the mentioned points but many others as well. Therefore, judging carefully the risk of releasing transgenic plants is a subtle issue also because the economic pressure and expectations of companies are heavily involved.
In practise, a responsible risk assessment of authorities often is hampered by the fact that companies do not disclose important information due to a conflict of interest. Sometimes also official guidelines for risk assessment do not include all points necessary for a comprehensive analysis and should frequently be reconsidered. Responsible controlling agencies and scientists involved into risk assessment and releasing transgenic plants should be always very careful and should feel very much responsible. The question of a moratorium of releasing transgenic plants seems to be debatable even though it would not change possible effects of transgenic plants already released like in cotton with the gene of the *Bacillus thuringiensis* toxin.

During the discussions the following questions were raised but there was little time to comprehensively elaborate on all of them. A summary of the main points discussed is given.

1) Does each gene call for one trait or more?

Tissue cultures and optimal lab growth may produce different results as naturally grown plants when adaptation and selection come in. When one gene has more than one phenotypic expression or when a phenotype is genetically controlled by more than one gene problems may occur.
Also genetically engineered plants have to undergo a selection process after they have been produced in the lab in tissue cultures. They have to be grown as ordinary plants in order to test the introduced trait is kept stable and really exhibits the desired phenotype. There is always the possibility that the trait gets lost, e.g. by recombination in later generations. Only if the trait is really stable the plant can be used further. There are examples that an introduced gene produces different results than expected (unexpected high incidence of changed colours in transgenic petunia at the first large-scale field experiment of the Max Planck Institute for Breeding Research in Cologne, Germany, in 1989) indicating that the relation between genotype and phenotype is complex or that the phenotypic expression of a trait depends on the environment.

2) Is it right that genetic engineering is just an acceleration of traditional breeding?

There was some controversial discussion on this point. On the one hand, genetic engineering can be used in a way of just accelerating the traditional breeding process.

On the other hand, genetic engineering can overcome barriers of traditional breeding by introducing genes and traits into a host from a source which was unavailable hitherto. The second point makes the application of genetic engineering in plant breeding so appealing.

3) Genetic engineering is assumed to be precise and classical breeding unprecise, why?

Classical breeding used to be and is mainly still today based on long-term breeding and selection processes without having a clear-cut insight into the genetic basis of the traits for which is selected. On the other hand, by the need of always growing the plants it becomes clear whether the trait of choice is selected in the right way or not. Hence there is always a strict empirical control of the breeding process.
Genetic engineering, by starting on the genotype level and being a late product of our cause-and-effect thinking, is usually applied for traits that are controlled by so called "major genes", i.e. genes that have by themselves a measurable influence on the trait at phenotype level. This does not mean that the trait is only controlled by the introduced gene(s). Whether the gene transfer has changed the trait at phenotypic level must be shown by growing the plant. This is the control whether the method and assumptions for introducing a given trait were correct. Therefore, even though genetic engineering is assumed to be precise, it is not as long as we do not have a comprehensive insight into all details of the highly complex plant organism and its environment of choice.

4) Do the high promises and expectations in gene technology by the industry hold true?

As far as we can judge today from the few examples of commercially growing transgenic plants the promises are at least questionable if not proven wrong. First, according to some information one participant brought in, the FlavrSavr tomato turned out to have a much lower yield than expected which made it very difficult so far to sell it. Second, gene transfer through hybridisation among species appears to occur with a much higher probability than expected. Every crop grown in an ecosystem can naturally be the source of a vertical or horizontal gene transfer to other organisms. The probability of outcrossing is given by the way of pollinating (e.g. rape = outcrossing species; wheat = self-pollinator) and the presence of closely related species in the neighbourhood of the crop. Genetic engineering by itself is not likely to change the frequency of neither vertical nor horizontal gene transfer. However the spectrum of genes brought in is enlarged compared to classical breeding. A gene transfer in nature is considered unwanted by most participants if it concerns a trait that affects the ecosystem, as for instance herbicide tolerance transferred to wild plants, e.g. from rape to other brassica species in Danish experiments in 1996. This fast vertical gene transfer seems to misprove the promise of reducing the application of herbicides. In the case of self-pollinators as wheat or crops without related species in the ecosystem as maize, the situation is different. One participant reported that the introduction of the gene coding for the *Bacillus thuringiensis* toxin into cotton had the unexpected and unwanted effect that not only the parasite larvae but also many other insect larvae were killed. Even though the companies claim that these unfortunate effects are just because we are still in the childhood of the application of genetic engineering in plant breeding the cited results strongly weaken the high expectations in this context.

5) Is it really true that transgenic plants will help to overcome the famine in the third world countries?

So far there is nothing than just pure hope that transgenic plants will really help to overcome the famine in the third world. As we all know hunger in the third world is mainly a problem of distributing the food. There is a huge overproduction of food in more developed countries which first should be used to overcome the hunger. It appears more that the argument of solving the problem of hunger in the third world is put forward by companies to help accept genetic engineering in plant breeding and to keep expectations high. The same argument was used in the sixties when the "green revolution" was introduced which, as we all know very well, did not solve the problem of hunger in the third world.
Transgenic plants with traits appropriate for a higher quality or more appropriate for growth in certain climatic regions may contribute as one of many means to reduce the hunger but certainly is not the means of choice. However, there is still the question of whether we really come to better plant varieties by genetic engineering or may be also by careful traditional breeding. So far, genetic engineering aims for few varieties which are used most widely. A meaningful traditional breeding applied in many regions of the world may come up much faster with better varieties in a given climate zone.

6) What are side effects of gene technology?

A general statement on side effects of transgenic plants is not possible. Every single case must be checked for side effects. Therefore the legislation in Europe provides that for every release of transgenic plants the applicant must present studies on possible side effects, such as changes in the toxicological or allergenic potential or the nutritional value (carbohydrates, fats, proteins, micronutrients, etc.). He must present data of the behaviour of the transgenic plant in the ecosystem (competition factors, reproduction) where the release is planned. This so called "case by case risk assessment" seems to be a reasonable way of handling possible side effects, not only of transgenic plants, but on transgenic organisms in general.

7) Why does the public often not accept gene technology and gene food but accepts other technologies as computers and cars?

It is not true that the public only gets uneasy about transgenic organisms an not about other technologies. Often when a new technology was introduced the public had a similar uneasy feeling about it. On the other hand the point which is different with genetic engineering and transgenic organisms as compared to other technologies is that no technology before had such a great potential to affect and manipulate life, and thus the destiny of man.

20 Intrinsic value of plants and animals: from philosophy to implementation

WORKSHOP

PETRAN KOCKELKOREN
Environmental philosopher
Department of Philosophy, University of Twente
PO Box 217, NL-7500 AE Enschede

MICHIEL LINSKENS
Biologist and policy advisor
Dutch Society for the Protection of Animals
PO Box 85980, NL-2508 CR The Hague

1 The intrinsic value: fundamentals

In March 1993, the Dutch Section of *If*gene took the initiative for a public debate in Amsterdam on the supposed benefits of biotechnology. The chairman of the NIABA, the Dutch Society of Biotechnological Engineers, took the opportunity to give a full exposition of his ethical stance. This can be summarized in a simple formula: biotechnology in relation to humans, 'No!' (only in rare exceptions, when lives depend on it in the medical sphere); biotechnology in relation to animals, 'No, unless' (unless, again, human lives are at stake and can only be saved by medicines of animal origin); biotechnology in relation to plants, 'Yes, provided that' (sufficient safety measures are met); biotechnology in relation to bacteria, 'Yes'.

The difference between 'No, unless' regarding animals and 'provided that' regarding plants, consists in the reversed burden of evidence. 'No, unless' means that biotechnological engineers may not tinker with animal genes unless they themselves give sufficient reasons to make an exception. 'Provided, that' means that biotechnological engineers may tinker with plant genes unless a third party for instance consumer organisations gives sufficient reason why in this particular case they may not do so. All in all, the proposed hierarchy in ethical deliberation regarding intervention in genes of respectively human, animal, plant and bacteriological origin, shows a flagrant anthropocentric bias.

The closer the life forms concerned are to human biological architecture the less intervention is allowed and vice versa. Man and his health is the ultimate measure in tinkering with nature.

Anthropocentric bias

It is not difficult to make this ethical stance ridiculous with a rhetorical appeal to the value of all life in general.

Man is then exposed as a crude and cruel ruler. That would perhaps be gratifying for nature lovers who congratulate themselves on their own sensitivity but it would certainly not be justified in the light of the history of ideas. Of course one can cry out that such an anthropocentric scale of values reduces other life forms to their functional value for human comfort, as if they don't matter in their own right. Indeed, one can put forward a solid argument that nature is then regarded as a mere provider of raw materials to be technically processed. Nature would only be ascribed an instrumental value.

Yet the chairman of the NIABA hastened to dismantle these awkward arguments. He appealed to a higher court, that of God, Who gave man a distinguished task in steering natural evolution under His supreme guidance. The chairman opted for the benevolent role of stewardship over nature. Look at the random drift of evolution, he said, look at the callous waste of lives. For every successful mutation, which just might be slightly better adapted to its ever changing environment than its competitors, multitudes of misfits are wasted, often at severe cost in terms of suffering. With the help of biotechnology we can make shortcuts in the evolutionary trajectory. We can rule out lots of misfits beforehand. We can improve on nature's blind meanderings. We should take up our responsibility instead of shrinking from it with pious prevarications. A shortcut on suffering is man's contribution to the history of nature. Nature can flourish better if man participates technically in her lavish productivity.

Even without God behind the scenes this argument holds at least some water. Still, it is not clear why the justification of any possible intervention should always be in favour of man. A rightful steward should not in all circumstances sacrifice the so called lesser life forms entrusted to him to his own welfare. One would like to have a more impartial standard of ethical deliberation than a biased hierarchy in which man places himself at the top, supported by a God who allegedly only favours him. Technological intervention might in some cases be defendable, but then it should be argued in terms of the welfare of the organisms concerned and not solely with a keen eye on human profits. But how can animal and plant wellbeing be conceptualized without having recourse to spurious notions such as never hurting the feelings of animals and plants, or regressing into antiquated essentialistic musings? As soon as one pretends to have privileged access to the inner essence of any life form, one promotes a dualistic epistemology as well. On the one hand there would be the 'real' nature of an animal or plant, on the other a scientifically reduced picture of it. Ethics is in the same movement transformed into therapy for scientists and engineers. The latter should accordingly learn to turn back to unprejudiced perception, to trust in their authentic preverbal relationship with nature. No wonder the chairman of the NIABA recoils in horror from this kind of divinatory self righteousness. A trained scientist aware of the shortcomings of his method and of the limited applicability of his models is to be preferred above an exalted intuitionist who pretends to have an immediate hotline to nature's inner recesses.

With these severe admonitions well stocked in the back of our minds, let us turn now to the articulation of an ethics that does not suffer from an anthropocentric bias; an ethics that may adequately be called ecocentric without us having to take recourse to some higher faculty that penetrates into the very core of creation. We can start with the elucidation of the notion of intrinsic value, as derived from the philosophy of Kant, who, regarding scientific clarity, is above all suspicion.

Intrinsic versus instrumental

Can the notion of 'intrinsic value' be brought into play as a redeeming concept over and against the idea of a merely instrumental or functional value? The notion of 'intrinsic value' was coined by Kant. By him it is however solely applied to humans. Always treat fellow human beings as goals in themselves, never reduce them to instrumental means for attaining your own goals, he said grosso modo. Many contemporary philosophers of nature try to extend this notion to animals and plants as well. They too should be treated as representing goals in themselves. But in what sense can this be said to be true? Animals and plants don't have explicit goals like we do. In trying to lend some philosophical credibility to this ethical position towards nature, philosophers often take recourse to the ancient or classical idea of 'entelechy'. This idea implies that every organism in nature is oriented to an immanent goal ('telos' means 'goal') in its development, which is to be realized in the process of growing. While in a causal interpretation of events one imagines a motivating force from behind so that every effect is preceded in the past by its cause entelechy can be best understood as a front wheel drive: evolution is pulled ahead by its anticipated completion in the future. So a caterpillar has no choice but to turn into a butterfly, an acorn develops inevitably into an oak, and so on. From this conception of the state of affairs it is but a small step to an ethical interpretation of natural processes. One should indeed respect this goal orientedness of all organisms and give other life forms than our own the chance to complete their full life cycle undisturbed. This doesn't mean that one may never eat a cauliflower but it does mean that one shouldn't tinker with cauliflowerness.

A new dimension

Within this philosophical line of reasoning, which links the concepts of intrinsic value and entelechy, it is possible to give a new dimension to the attitude of stewardship over nature. The golden rule: 'never exceed the boundaries of an organisms goal orientedness' doesn't ethically preclude technical interventions in the process of the organism's self unfolding. Technical intervention should go along with nature in her own processes, so that she can flourish in man's company and custody if need be. With due respect one should never transgress upon nature's ability for self regeneration. One may technically support nature's ways but one should not exceed what nature provides by forcing new goals into the game. This doesn't exclude biotechnological interventions in principle, but it does demand that arguments for doing so are stated in terms of nature's own direction.

Although it may seem that by having recourse to the classical view of nature we have found a viable ecocentric alternative to the anthropocentric position, critics in the latter camp are not silenced. The classical conception of intrinsic value is met with serious objections. One must have a very harmonious view of nature to endorse the notion of entelechy. It is as if the boundaries of 'genus and species' are fixed forever in a one track evolutionary scheme; as if all life forms in their different grades of complexity can be stacked in a neat pile or pyramid of concepts with God smiling at the top. Don't dare disturb the beauty of this giant ant heap. The chairman of the NIABA, on the contrary, looks at evolution rather as a random process accompanied by myriads of casualties alongside the royal road. If one intervenes technically in the process of evolution it is very difficult to make out where the boundaries lie. "Genus and species" are fuzzy concepts. The life forms demarcated thereby are involved in a continuous process of trial and error wherein the boundaries themselves are reshaped over and over again. There are no clear cut lines in sight that are to be respected. The anthropocentrists may have an irrational belief in the nobility and superiority of man but the opposite position is not less irrationally exalted over the fancied self regulatory harmony in nature. Yet we can hang on to the notion of intrinsic value in nature without having to subscribe to an over harmonious view of it. Let us inspect the evidence. Every year in spring all life forms multiply fivefold: five times as many seagulls, foxes, frogs, whatever. In autumn we end up with the number it all started from. This means that for every singing bird, four others die, often gruesomely; ripped to pieces and swallowed by the cat, smashed by falling out of the nest, flattened out on the road surface by passing cars, eaten from the inside by parasites, and so on. Nature is very wasteful with individuals in pursuing its lofty goals. We should realize that we are in this predicament all together, sharing the same vulnerability: plants, animals, humans. Nature is less partial than we are in bestowing blessings and curses. All the different life forms are rowing against the stream or even against cascades of deterioration, all obedient to the law of entropy, resulting in everlasting dissipation of energy.

Life is counteracting this law. Entelechy is not pre-given as if it is never contested. There is no pre existing order. Order is being created in actu, in an actively woven interdependence of life forms hovering above the abyss of decay. Human ethics with regard to nature is a question of voluntarily shared vulnerability. That is why one should respect the full completion of the life cycle of any life form, even if its boundaries are not clearly discernable.

But what kind of restrictions does nature impose upon technical interventions in its self ordering processes? Well, just because nature's goals are not pre-given in blueprints from the beginning of time, we are not allowed to reduce nature's workings to the metaphor of the genetic code. Goals are not pre-given but sought out and realized in the process, every life form being an extrapolated tentacle of nature's probings into the unknown. Of course there is a genetic code that fixes a record of past endeavours. But the process itself is only actualized in the interaction of the genetic code and the environment which is conditional for its expression. There is no evolution of genes, only a coevolution of organisms and environments (including other organisms) and the outcome of this is registered in genetic codes. Only in the interaction and interweaving of genes as evolutionary mnemonic devices on the one hand and the material environment on the other can the bold venture of evolution take place.

In this perspective biotechnology is not a shortcut in the shunting yard of evolution, rather it suspends the whole risky enterprise of interlocking life forms completely. To be successful, biotechnology has to standardize the conditions of expression for the modified genes in question. The precarious material conditions are best eliminated all together, as in hydroponics or in mineral substrate cultivation. The vulnerable exchange between life forms is replaced by the forced expression of isolated genes in universal media. An ethical protest against these practices has nothing to do with a sentimental yearning for a paradisiacal state that never existed anyway, but it does appeal to human solidarity with our fellow life forms that, given half a chance, share our jeopardy with us.

Looking back upon our efforts so far, we can say that we have scanned a foursome of basic attitudes over and against nature. Ethics draws its validity from one of these ideal typical frameworks. The first we discussed is the attitude of rulership over nature, which finds its justification in the view of random evolution. Nature is a malicious game of roulette under the sole supervision of man. The second attitude is that of stewardship, which may be understood as a form of rulership mitigated by compassion. Man forms the pinnacle of evolution and, since 'nobility obliges', man makes technological shortcuts in the unpredictable ramblings of evolution. On the other side of the spectrum, diametrically opposed to anthropocentrism, stands ecocentrism, which finds its supreme justification in the classical theory of entelechy in nature. This theory condemns us to a sheer participating fatalism in the end, because only meek handcraft technologies meet the demands of a harmonious goal oriented nature. There is a fourth position in between the extremes of dominating anthropocentrism and participating ecocentrism: that of partnership in which nature contests playfully with human steering capacities. Our hope for the future rests on the assumption that we are witnessing a societal shift towards this attitude of partnership with nature!

2 The intrinsic value of animals: implementation into legislation

2.1 Introduction

The intrinsic value of animals is since twenty years an important issue in the Netherlands. The term was introduced to give counterweight to the instrumental use of animals in for example factory farming. In the past twenty years many participants were involved in the debate on the recognition of the intrinsic value, like politicians, philosophers, scientists, animal welfarists, civil servants. Finally the recognition of the intrinsic value was implemented in Dutch legislation in 1992 and 1996. In this paper the history will be described, the implementation in two important animal welfare laws and how the principle of intrinsic value was translated into a set of tools for ethical review and governmental decision making.

2.2 History

In 1981 the report "Government and Animal Protection" was published by the Dutch government. In this report specific attention was paid to the intrinsic value of animals. It was stated that "the animal protection policy in the Netherlands should be developed based on the recognition of the intrinsic value of the animal". Furthermore it was said that "the policy should be focused on an optimal protection of the animal against human activities which may inflict his physiological and ethological welfare. The interest of the animal has to be incorporated in a decision making process". These wordings were very important for the debate which started soon afterwards on the development of new animal welfare legislation. It took over ten years to finalize the discussions and to publish the new law on animal welfare.

Another important discussion took place because of the rapid developments on biotechnology and animals. Many participants in the debate on the new animal welfare legislation agreed that biotechnology on animals (including genetic engineering and cloning) was one of the most important threats to the intrinsic value of animals. In 1989 the government installed a specific committee to prepare and advise on this important issue. The commission was chaired by prof. E. Schroten. In 1990 the commission published the report "Ethics and animal biotechnology", in which six principles (see box) were introduced, based upon the recognition of the intrinsic value of animals. Besides normative principles, like 'respect for integrity' and 'to do good' also two procedural principles were introduced. The first one, the principle of control, focuses on the necessity of a democratic process, in which all actors should be able to present their opinion. The new legislation on biotechnology on animals (will come into force in april 1997) is part of the new Animal welfare law and has implemented this principle in the procedure. The second principle, the principle of correction, implies that the engineering of the animal should be revocable. In most cases it will be possible to 'revoke' animals. In some cases however it seems difficult to 'revoke' animals, like escaped transgenic mice or transgenic fish which are deliberately released in the environment. For those cases specific constraints are introduced.

Principles used in ethical review

1 Respect for integrity
2 To do good (care for animal health and welfare)
3 Refrain from doing harm (not harm animals)
4 Justify (avoid needless damage and risks)
5 Control (democratic process)
6 Correction (revocability)

2.3 Implementation into legislation

In september 1992 a new *Animal welfare law* in the Netherlands was accepted. This law is based on the principle that the intrinsic value of animals is recognized. Due to this principle the legislation on biotechnology on animals has a "No, unless" policy. This implies that genetic engineering of animals is forbidden, unless:

- the action do not have unacceptable consequences for the health or welfare of the animals;
- there are no ethical objections against that actions.

Related to those ethical objections several criteria are introduced for an ethical review on every experiment with genetic engineering of animals (see box). The criteria reflect the recognition of the intrinsic value and are derived from the work of the ethical committee in 1990. The most important criteria, besides the animal welfare criterium, are that the goal should be necessary (e.g. to production of live saving drugs) and that no alternatives should be available (see further).

Criteria in ethical review

1 Substantial interests
2 Alternatives available
3 Harm to the animal

The recognition of the intrinsic value is also integrated in the revised *Law on laboratory animals* (1996). In this law a specific article is introduced on intrinsic value:
"In practising this legislation the recognition of the intrinsic value of the animal is accepted as general principle".

2.5 Animal welfare law: ethical review

Every experiment in which transgenic animals are developed has to be ethically reviewed. The ethical review will be conducted by a specific committee with experts in the field of biotechnology, ethics, human health care and animal welfare. This committee has to advise the ministry of Agriculture on every project with genetic engineering of animals. The minister of Agriculture will finally decide if the project will get permission to start.

In the ethical review the ethical commission is using a two-step approach (see box). First three criteria should be met before the second step in the procedure will be made.

Substantial interests (step 1)
In the first place the experiment should serve substantial interests. It is not very clear yet which interests will be considered as 'substantial'. An experiment like the use of transgenic animals for studying human diseases or the use of transgenic animals for the production of (life saving) drugs will be considered as 'substantial'. It is questionable if an experiment with transgenic livestock for the production of human proteins used as nutritional food may be considered as substantial.

Ethical review: two step approach

Step 1
1 Substantial interests
2 No alternatives available
3 No unacceptable harm to the animal

Step 2
Positive conclusion on three criteria, then consider:
4 Expected and unintended consequences
 * to the animal
 * to the health and welfare of human beings
 * to nature
 * to the environment
 * to society.

Alternatives (step 1)
In the discussion on genetic engineering of animals the availability and possibility of alternative production systems is an important issue in decision-making. From an ethical point of view it is necessary to be able to choose between several options. Without an alternative option it is very easy for technologists to get their option accepted by the public.
Considering the different applications of genetic engineering of animals the production of human proteins in milk/blood of genetically engineered animals has to deal with

several very strong alternatives. The strongest alternative option for the production of human proteins is the use of genetically engineered micro-organisms, like yeast, bacteria and fungi (Aspergillus). Many proteins already are produced on large scale by these organisms, like human insulin, bovine chymosin, human clottingfactor VIII and human lactoferrin.

Animal suffering (step 1)

Thirdly the ethical committee will take into account the harm to the animal. The technology is still in an early stage of development, although transgenic animals already exist since the early eighties. Due to the state of the technology and despite the experience with transgenic animals many animal welfare problems still occur. Animal welfare implications are already reported concerning donor and recipient animals, laboratory animals and farm animals. It is only since 1994 that some research (in the Netherlands) has been set up to investigate animal welfare problems.

Donor animals are subjected to hormone injection, surgical insemination and surgical removal of embryos. A new technology to collect high quantities of eggs from living animals has been recently developed (Ovum Pick Up method). Every two weeks twenty ore more eggs are collected from the ovary of cows with a big tube. Recipient animals are also subjected to hormone injections and surgical implantation of embryos.

Animals used as a tool for modelling human diseases, like cancer, AIDS, etc. are suffering to a great extent. Those experiments, including the horrible suffering, are legitimized by society because of the value for treating patients and the improvement of human medicine. Besides suffering caused by the inserted gene, also suffering is caused by technological failure.

In the development of transgenic livestock several animal welfare problems are already reported (O'Brien, 1995), like the incorporation of growth hormone genes in sheep, which has resulted in disrupted joint development, diabetes-like conditions, and reduction in longevity. Another example are the pigs in Beltsville (US) in which growth hormone genes are incorporated. This experiment has resulted in gastric ulcers, liver and kidney damage, bone and joint problems, loss of coordination, sensitivity to pneumonia, damaged vision and diabetes-like conditions.

It is very difficult to estimate animal welfare problems in future because of genetic engineering. However it is plausible that an increase in applications will lead to animal suffering. For example the effect of gene transfer may not be apparent in the immediate progeny. Undesirable effects may only come to light in later generations, when the transferred gene combines with other part of the recipients genome in an unforeseen way. Another animal welfare problem which may rise is in the production of pharmaceutical proteins and human organs in transgenic livestock. This will be accompanied by severe measures to prevent contamination with unknown viruses, like BSE and scrapie. It is expected that those animals will be kept individually in sterile conditions (specific pathogen free, SPF), without possibilities to go outside. It is estimated that hundreds of thousands of pigs are necessary to produce for example human haemoglobin in their blood. Thousands of transgenic animals are necessary to produce enough quantity of pharmaceutical proteins or 'human' organs.

Consequences (step 2)

If the answer on the first three criteria is positive, then another criterium has to be considered, namely the expected and unintended consequences. This criterium does not focus on the animal itself (as the other three criteria do) but on the consequences of the activity on other actors or issues, like nature and the environment. It is expected that one or more negative consequences will lead to an extra set of constraints to cope with those consequences (like to effects of transgenic fish on the environment).

Considering the above mentioned aspects on the three criteria and the consequences it may be clear that the first year after the legislation has been set into force (april 1997) it need extra effort to clarify the three criteria. It should be clear to all actors involved what is meant by substantial interests, how far an alternative should be developed (should it be ready for use or in development?) and how animal suffering (now and in the future) will be weighed.

2.6 Animal welfare law: questionnaire

The ethical committee developed questions which should be answered by those who are willing to start research on genetic engineering with animals. The next questions should be answered:

General

- What is the goal of the project (short term/long)?
- What is the importance of the project (necessity)?
- What are the interests involved?
- What are the expectations about successfulness?
- Which expertise is available?

Specific to the project

- Which technique is used?
- Where does the material come from?
- What is the number of animals used?
- Which species?
- How the animals will be housed?
- What is the destination of the animals/material after the project?
- What are unintended consequences?

– What are positive and negative effects on the health and welfare of the animals?
– Are alternatives available?

The dossier from the scientists should include the answers on these questions and will be used by the ethical committee as a starting point in the process. If some questions are not answered properly, the committee will request for additional information.

2.7 Discussion

The participants of the workshop discussed on the most important questions to the scientists. The following questions were raised:

– What is the goal to genetic engineering with an animal? Is it for human purposes (profit, luxury, health, agriculture) or in the interest of the animal?
– What about the quality of the product?
– Are there any alternatives?
– Is there a justification?
– What about animals suffering? Is it a question of the individual or of the species? Are there species-borders? Are "natural animals" involved in the scope of this question?
– What input has the concept of archetypal ideas of animals?

It is clear that the major questions of the participants correlate very well with the questions of the ethical review. The only question which will not be dealt with in the Dutch system is the fundamental question about the concept of archetypes. Before the legislation came into force in the Netherlands an extended discussion has taken place about the intrinsic value of the animal, in which a discussion on the context of archetypes would have been fit very well. For the Dutch system it seems now too late to raise this fundamental issue.

The participants also formulated a set of criteria. Next criteria were mentioned:

1 The animal may not suffer. Suffering means:
  – physical and physiological deformation (individual suffering)
  – improper habitat (ecological suffering)
  – unsuitable behaviour (ethological suffering)
  – spiritual suffering with regard to the archetype-concept
2 The integrity of a species has to be respected:
  – Are the boundaries of species crossed?
  – Are there effects on the offspring?
  – Is there a loss of instincts?
3 Are there effects on the environment?
4 Are there any alternatives?

These criteria also meet very well the criteria used in the two step approach in the Netherlands. The issues mentioned on animal suffering by the participants may be of use for the implementation of the criteria in the Netherlands. It was made clear that due to complexity of the criteria it is very important that within one year a clear description of each criterium have to be developed. The remarks of the participants on the integrity of species is probably the least recognizable in the Dutch system.

An important remark is that the introduction of the intrinsic value in the Netherlands caused a change in moral intuition in society to treat animals with more respect. It depends on the specific implementation of the intrinsic value in legislation to come to a final judgement if animals will be better off.

In this workshop it became clear that in the Netherlands the fundamental discussion on the recognition of the intrinsic value has been replaced by a pragmatic way to implement the concept of intrinsic value in legislation and political decision making. On the other hand, most participants (not from the Netherlands) were interested in fundamental questions, like "How to find ethical values for our activities as human beings, for our research, and especially for genetic engineering"? Those questions were discussed in the workshop and may be of great importance to the discussion in the European Union and the member states on specific legislation on biotechnology and animals.

References

Advisory Committee Ethics and Biotechnology in Animals (1990): *Ethics and Biotechnology*, The Hague.

Dutch government (1996): *Animal Health and Welfare Act.* The Hague.

Kockelkoren, P.J.H. (1995) Ethical Aspects of Plant Biotechnology, Report for the Dutch Government Commission on Ethical Aspects of Biotechnology in Plants, in *Agriculture and Spirituality, Essays from the Crossroads Conference at Wageningen Agricultural University*, International Books, Utrecht.

Langley, G. (1995). *Genetic engineering of laboratory animals – causes of suffering.* On behalf of the British Union for the Abolition of Vivisection, London, UK.

NOTA (1990) *The created animal (Ethics and genetic engineering of animals.* Summary. NOTA, The Hague.

O'Brien (1995). *Gene Transfer and the welfare of farm animals, Compassion in World Farming*, Petersfield, UK.

Verhoog, H. (1995): Gentechnik und die Eigenwürde der Natur (Mein Weg zur ethischen Urteilsbildung. *Elemente der Naturwissenschaft* 63 (2): pp. 1-13

Verhoog, H. (1996): Genetic Modification of Animals: Should Science and Ethics Be Integrated?. *The Monist* 79 (2): pp. 247-263

Acknowledgement
We thank Hans-Christian Zehnter (CH – Dornach) for making the report on the discussion.

21 Does human retardation occur at the molecular level?

WORKSHOP

JOS VERHULST
Morphologist
Karel Oomsstraat 57, B-2018 Antwerp

NICOLAAS G.J. JASPERS
Molecular geneticist
Department of Cell Biology & Genetics, Erasmus University Rotterdam,
PO Box 1738, NL-3000 DR Rotterdam

Abstract
Homo sapiens is characterized by a very generalized anatomy. On the one hand, human morphology shows many occurrences of neoteny. On the other hand, many developmental processes in humans can be understood as straightforward extensions of generalized developmental patterns, that are expressed to a lesser extend in non-human mammals (hypermorphosis). These observations at the macroscopic level can be expected to correlate with certain particularities at the molecular evolution. Examples presented here are the so called 'hominoid slowdown' and 'introns early' hypotheses of molecular evolution.

Introduction
From 1918 onwards, Louis Bolk formulated his so-called retardation theory. According to this view, humans are characterized by retarded or arrested development with respect to other primates, or mammals generally. For instance, the hair coat develops in all primates according to a craniodistal pattern, with scalp hair appearing first. In humans, the development seems to be arrested at an early stage, and the hair coat on the trunk and limbs does not fully develop (for other examples, see Verhulst 1993, 1994). Bolk linked this type of neotenic phenomena with another observation, namely the very extended lifetime of humans. Human development is very slow, as compared to the development of non-human mammals of similar weight. Bolk supposed that this

ontogenetic slowdown resulted in the elimination, in humans, of final stages in animal development.

In the original version, Bolk's theory met with two problems. Firstly, it was not clear how retardation could lead to the final elimination of traits (delay does not imply deletion). Secondly, Bolk was confronted with some embarrassing 'exceptions'. For instance, both humans and non-human primates are born with short legs. Thereafter, humans develop very long legs, the hindlimbs of non-human primates remaining much shorter. The long legs of adult humans are instances of hypermorphosis. Bolk considered these 'consecutive' phenomena as genuine examples of adaptation in the darwinian sense. The core of the hominization process he saw in the process of fetalization, that was (in his eyes) not explainable in darwinian terms.

Recently, a new version of the Bolkian theory was proposed (Verhulst 1993, 1996, 1997). The core concept is that of a generalized ontogeny, a developmental pattern shared by humans and non-human primates or mammals. This developmental pattern is thought of as a bunch of abstract developmental paths, each path consisting of a large number of subsequent developmental steps.

This generalized pattern does not encompass any adaptation to specific ecological niches. In order to insert such adaptations, parts of the generalized ontogeny have to be eliminated and replaced by specialized developmental steps. As a null hypothesis, we accept that the chance of a generalized developmental path being interrupted to permit replacement by a specialized developmental sequence, is about equal at every stage of that path. It follows, that steps in the later parts of the generalized ontogeny are increasingly likely to become eliminated by specialization. In other words, the later stages of the generalized ontogeny are less likely to survive in a specialized animal in the ontogeny of the specialized animal, the early developmental stages will be more general, the later stages will be increasingly specialized. This conclusion is nothing else than the well-known 'law of von Baer'.

One further axiom is introduced: in order to insert a certain number of specialized developmental steps, we have to eliminate a greater number of original, generalized steps. This is because the original developmental scheme forms a coherent whole, that is disrupted by external needs when specialization occurs. One can compare this circumstance with a tower, that is planned and constructed to attain maximal height. When the plans are changed during construction, the optimal height will not be reached, because the coherence of the original plans has been interfered with. We thus expect an inverse relationship between lifetime an degree of specialization or adaptation of a mammal. The more specialized a mammal, the shorter its lifetime will be.

This new version of Bolk's theory explains why primates, that show on the whole less specialization as compared to other mammalian orders, also tend to have longer lifetimes. Humans show very few specializations, and have an exceptionally long lifetime. According to this view, Homo sapiens is the truest expression of the generalized ontogeny of mammals.

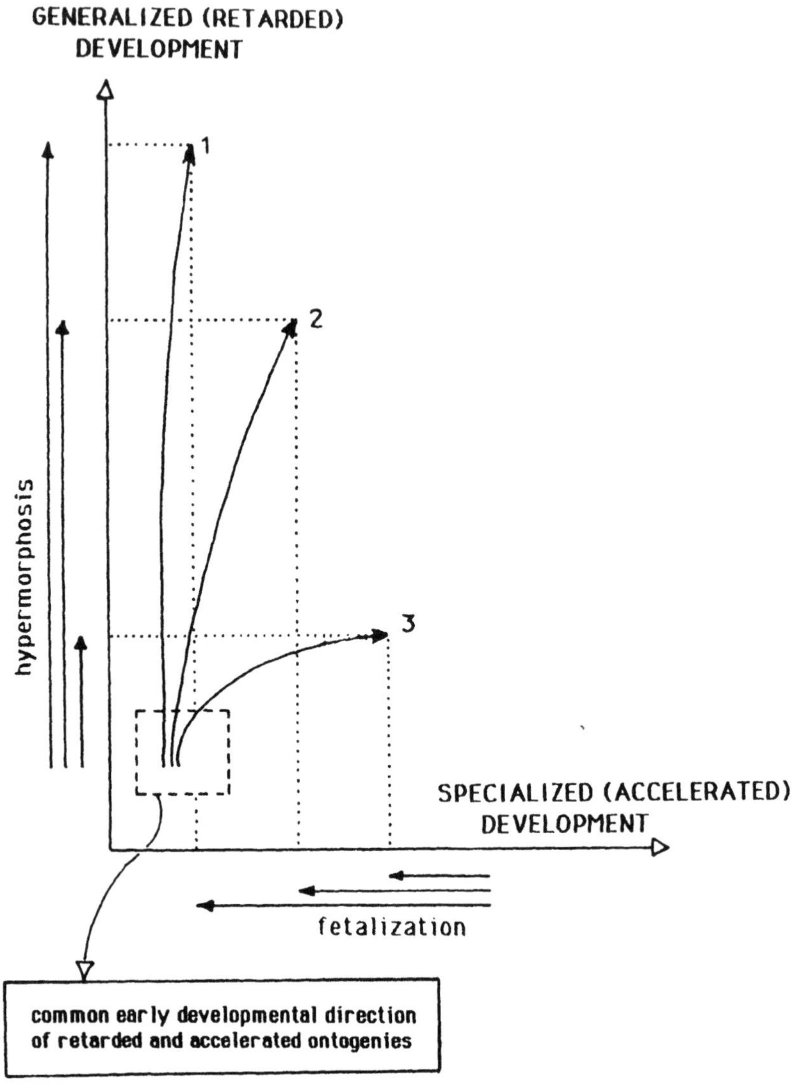

Figure 1. Hypermorphosis and specialisation

1: very retarded, humanlike life history; 2: moderately retarded life history (for instance of an ape); 3: accelerated life history of a typical mammal. Longer histories are represented by longer arrows. The ontogenies develop along two complementary dimensions. The vertical dimension represents the degree of expression of generalized ontogeny and of hypermorphosis. In more retarded ontogenies (such as 1), the anlagen of the generalized ontogeny are developed more consequently and thoroughly. This will result in a comparatively stronger development of late-coming organs and traits. In more accelerated ontogenies (such as 3), the late-coming organs and traits will remain comparatively smaller, the overall life history will become shorter and the degree of specialization will be greater.

As a matter of fact, this expanded version of Bolk's theory was already summarized in Goethe's diction: 'On the one hand, animal-like traits are suppressed in humans. But on the other hand, animal development moves up to a higher level in humans'. That is to say: in humans the generalized Bauplan reaches its most consequent expression, whereas at the same time, specializations and adaptations remain minimal. Fig. 1 summarizes this concept.

A slowdown of the molecular clock in the human lineage?

The molecular clock hypothesis was proposed by Zuckerlandt and Pauling in 1965. It states that the rate of molecular evolution is approximately constant over time and along different lineages, thus permitting the establishment of evolutionary trees.

Even before this thesis was put forward explicitly, Goodman challenged the molecular clock hypothesis and suggested that the rate of molecular evolution slowed down along the evolutionary lineage of humans (see Goodman 1996 for references). At that time, in the early sixties, the evidence came from immunological data; but later, studies of protein and DNA sequences lead to the same conclusion. Goodman and other researchers think that "...the hominoid slowdown in rates can be attributed largely to decreasing mutations rates resulting from lengthening generation times and the evolution of more efficient DNA repair mechanisms" (Goodman 1996).

The hypothesis of an hominoid slowdown did not receive unanimous acceptation. For instance, Sarich and Wilson attributed the small immunological distances not to a slowdown in rates of molecular evolution, but rather to late ancestral separation within the Hominoidea. However, Goodman argues that relatively late divergence dates still require an evolutionary slowdown: "Molecular clock calculations applied to amino acid sequence data placed the human-chimpanzee split at only 1-1.5 million years ago [...] When the dates used for branching points in the phylogenetic trees for homologous amino acid sequences were constrained by time limits indicated by the fossil record, rates of protein evolution were found to be much faster in the stem-eutherian to stem-anthropoid linkage than in later anthropoid lineages, the slowest rates being in the hominoid lineages" (Goodman 1985, 1996; Goodman *et al* 1990).

According to some researchers, the hominoid slowdown even discriminates humans with respect to the great apes. For instance, Li and Tanimura (1987) not only assert that "...the average synonymous rate in rodents is 4-10 times higher than that in higher primates, 3-6 times higher than that in monkeys, and 2-4 times higher than that in artiodactyls", but they add that "...the rate in the human lineage seems to be lower than those in the ape lineages. In all the comparisons between an ape lineage and the human lineage, there is no case in which the human lineage has evolved faster. This is true for both nuclear and mitochondrial sequences. When all the nuclear sequences are considered together, the rates in the orangutan, gorilla and chimpanzee lineages are, respectively, 1.3, 1.9 and 1.6 times faster than the rate in the human lineage".

Early evolution

The observation of the hominoid slowdown seems to remain restricted to the domain of the primates or mammals. But at the morphological level, the principle of retardation can be followed still further down the evolutionary tree (Verhulst, in press). For instance, within the chordates we can consider the vertebrates as less evolved with respect to the tunicates (as an example, the brain and spinal cord in tunicate larvae still have the overall constitution conserved in adult vertebrates). So we can expect that a retardation principle could operate at a larger scale as well. The question whether molecular-genetic retardation phenomena, can also be found over wider ranges of organisms will be the subject of the remaining sections. Nucleotide sequences of genomes from a wide diversity of organisms are becoming available nowadays with continuously increasing speed, so the amount of sequence information should be sufficient to obtain clues in this direction (Leipe 1996)

The most conspicuous overall genomic differences between organisms concern sequence, size and the ratio between coding and non-coding sequences. The conservation of nucleotide sequence (and the deduced polypeptide sequence) between corresponding genes of the species is significant enough to allow construction of elaborate evolutionary trees, usually fitting perfectly with the palaeontological data. Non-coding sequences on the other hand, differ widely: they are virtually absent in the procaryotes and make up around 90% of the genomes of higher eucaryotes including man. The conservation of these so called 'intervening sequences' (introns) is very low. In fact, they are so highly individualized in man, that they can be exploited as personal fingerprints, as has now become common practice in forensic genetics. It seems fair to state, that the relatively conserved coding sequences (only a few changes between man and apes) determine gross body plan and morphology of the species, whereas the non-conserved intervening sequences appear to correlate with inter-individual variation. It turns out that individualisation of non-coding DNA is the highest in man, where mating practice approaches the random levels more than in any other higher species.

The origin of introns

Soon after the discovery, in the late seventies, of non-coding intervening sequences (introns) in the genomes of eucaryotes, theories describing their origin were formulated. Two opposing hypotheses have been put forward in this respect. The controversy between the two, known as the 'intron-debate' has not yet been fully settled. The so called 'intron-early' hypothesis (also called 'the exon theory of genes') originally introduced by a number of outstanding geneticists (Gilbert, Blake, Doolittle, Darnell: all 1978) assumes that intervening sequences have already been present before the primordial split between eucaryotes and procaryotes. These introns, self-splicing initially, were lost in the procaryotes in later evolution in a genome streamlining process. In the eucaryotes the introns were retained, with the concomitant emergence of an efficient

RNA splicing apparatus and an organizational separation between transcription (= mRNA synthesis) and translation (= protein synthesis).

The opposing 'introns late' hypothesis (Rogers 1985; Palmer and Logsdon 1991; Cavalier-Smith, 1991) states that introns have started to appear only after the procaryote-eucaryote diversification, as a byproduct of so called transposable elements, molecular parasites of the genome of procaryotic (or viral) origin, which carry their own instructions for insertion and removal (Cavalier-Smith 1985). They could firmly settle in the genome after the transcription-translation separation in eucaryotes. The organism has to spend considerable energy to maintain these vast amounts of 'junk' or 'selfish' sequences; their ability to modulate the expression of the coding areas and the fact that they enable modular exchange of whole gene segments to speed up evolution are believed to be the main reasons for their continued existence.

A detailed discussion of the 'intron debate' is not possible within the frame of this workshop account. However, some salient points should be highlighted. Arguments in support of the 'intron early' option largely come from three sides. First, the (almost) exact positioning of introns in the coding sequences of widely diverging eucaryotes strongly favour a common origin: "lightning does not strike twice at the same spot". Evidently, this evidence must concern 'ancient' genes present in all organisms (Gilbert et al 1986). A few genes encoding glycolytic enzymes, such as triose phosphate isomerase (TPI) and glyceraldehyde-3-phosphate dehydrogenase (GAPDH) have since then dominated the discussion (Stoltzfus 1994). Due to an apparently quite extensive primordial intron loss even in higher organisms, the position of most introns (> 95%) as they appear in the present species (I remind the reader that sufficiently intact palaeobiological DNA is not available!) is equally well compatible with the 'late' hypothesis. Recently, even the most telling examples of intron coincidence in TPI and GAPDH were seriously challenged (Kersenach et al 1994; Kwiatowski et al 1995; Logsdon et al 1994, 1995). The second main argument for the 'exon theory of genes' is the assumption that genes split by introns serve evolution. New functions and proteins could quickly emerge by simple exchange of functional units ('mini-genes'), which could subsequently diversify to adaptation of new function (Gilbert and Glynias, 1993). The prediction would be that (i) traces of this type of 'exon shuffling' should be picked up in species and (ii) that present introns represent or correspond to functional domains in proteins (Patthy 1991). There is consensus on the existence of highly defined and conserved functional domains in even distantly related proteins, but conserved exon-patterning in these domains is not really apparent (Doolittle 1995). Compelling cases of evidence in favour of exon shuffling have been presented however (Long et al 1996), but most of these concern cases that occurred after the eucaryote/procaryote split. Another argument in favour of exon-shuffling is the fact that most exons occur in phase 1 of the reading frame, i.e. introns interrupts occur mostly between single codons and not in the middle of them (Long et al 1995). Thirdly, some recent support was found for early

introns on the basis of GC contents observed over the species (Forsdyke 1996).
Strong arguments in favour of the 'introns late' hypothesis have been obtained by numerous studies on transposable and movable elements, found in all eucaryotic species so far. The main body of the present evidence tends to lean over to the 'introns late' hypothesis (see also Mattick 1994), but two important points should not be neglected in this still controversial issue. First, most of the sequences have drifted to such an extent, that clear statements in favour or against either of the two are no longer easy to obtain: usually extensive statistical analysis involving many sequences is required to obtain just significant trends: exceptions are barely outnumbered by regulars. Secondly, and more importantly, it should be emphasized that both theories are not at all mutually exclusive. The most likely evolutionary scenario will probably include elements of both of them.
In the present context, we therefore suggest a scenario where the existence of primordial introns is likely. Some of them have been retained in eucaryotes, but also many have been lost in a continuous streamlining process and considerable numbers have been added as well, presumably by 'selfish' mobile elements for a considerable part, and also by other recombination events, especially after the emergence of an efficient splicing apparatus. For a large part these rearrangements may have been responsible for the ever increasing size of introns further on the evolutionary ladder. Exon shuffling has played a significant role in the evolution of new protein species; this role could even be extended after the eucaryote/procaryote division; the best signatures of such events only stem from relatively recent times after this division. In addition, recombinational processes have promoted in gene duplication. Newly duplicated genes have then diversified to distinct subfunctions (Sidow 1996). These events are the basis of the greatly increased genome sizes of higher organisms and have left their signature in the form of numerous so called 'gene families'.

Early introns as a molecular retardation phenomenon

The 'streamlining process' of the primordial genomes needs special attention here. While intermediate to low numbers of introns are still found in lower eucaryotes, they are almost absent in the more primitive eucaryotic species. In bacteria, there are none at all (at least in protein encoding genes). Moreover, the space between individual genes has been much reduced: in some cases adjacent gene codes have even been found to partially overlap. In small bacteriophages, where streamlining has reached its summit, examples exist of triple use of sequences in different reading frames. Similar to the phages, parasitic bacteria have also dismissed many of their genes, with the recently fully sequenced Mycoplasma Genitalium, carrying no more than 700 genes, as an example. This streamlining, which is evidently a process of elimination, is expected to result in a reduction in evolving power. In other words, genome streamlining is an expression of specialization. The most ancient and least complicated species that we know today, have thus carried this specialization process to the furthest end, assuming that the

streamlining process proceeds continuously. This specialization (i.e. lack of adaptive power) is probably the cause of extinction of most of the ancient species during evolution; still the bacteria can definitely be considered the most successful taxon on earth, if one considers the number of species and individuals present. With their relatively small and specialized genomes, bacteria have managed to colonize even the harshest places on earth, such as sites with boiling temperature and 3000 atm pressure!
On the other end of the spectrum, the mammals (including man) appear to have retained a large part of the original diversity in their genomes – in fact they have even added considerable heterogeneity. In the original terminology of Bolk's hypothesis this phenomenon appears a striking example of retardation. The idea is further illustrated in Fig. 2. The scheme highlights a hypothetical (minor?, heterogeneous?) subpopulation that we can no longer detect in the palaeobiological layers, which should have acted as an evolving reservoir. From this mother-line each presently known species could have branched off into specialization, by starting a process of genomic streamlining, which tends to 'fix' the species into their final appearance. Thus, species that have emerged most recently, have a relatively short history of specialisation: in Bolkian terminology, they can be considered as 'retarded', having conserved complex multipurpose genomes for the longest time.

Discussion: Implications of 'molecular retardation'

Because both concepts of a consequent evolutionary slowdown along the human lineage and the 'exon theory of genes' are still under debate, we cannot draw any definite conclusions. However, it is evident that these phenomena, if real, could receive a very natural explanation from a Bolkian viewpoint: the molecular slowdown could be nothing else but the reflection, at the molecular level, of the retardation at the morphological level. Just as humans show, as compared to other primates or mammals, a more retarded or fetalized morphology, they could also possess a less evolved DNA. This means that humans, as compared to apes or primates and mammals generally, remain closer to the common evolutionary stem: they retain a more generalized or less evolved morphology, and they also have less evolved DNA.
The question whether the retardation principle in the form of intron evolution can also be discovered in the more recently evolved species, which are addressed by most of the morphological studies, remains to be answered as well. In this respect, is interesting to note that some very clear examples of selective intron loss and exon shuffling have been described in vertebrate development (i.c. blood clotting proteins: Blake et al 1987).
From a more fundamental point of view, the Bolkian retardation theory seems to contrast with molecular reductionistic explanations of the origins of life, since a considerable degree of complexity is postulated to be present already from the beginning. Our attempt to describe molecular retardation phenomena may therefore, at fist glance, seem a *contradictio in terminis*. Our intention was merely to pick up

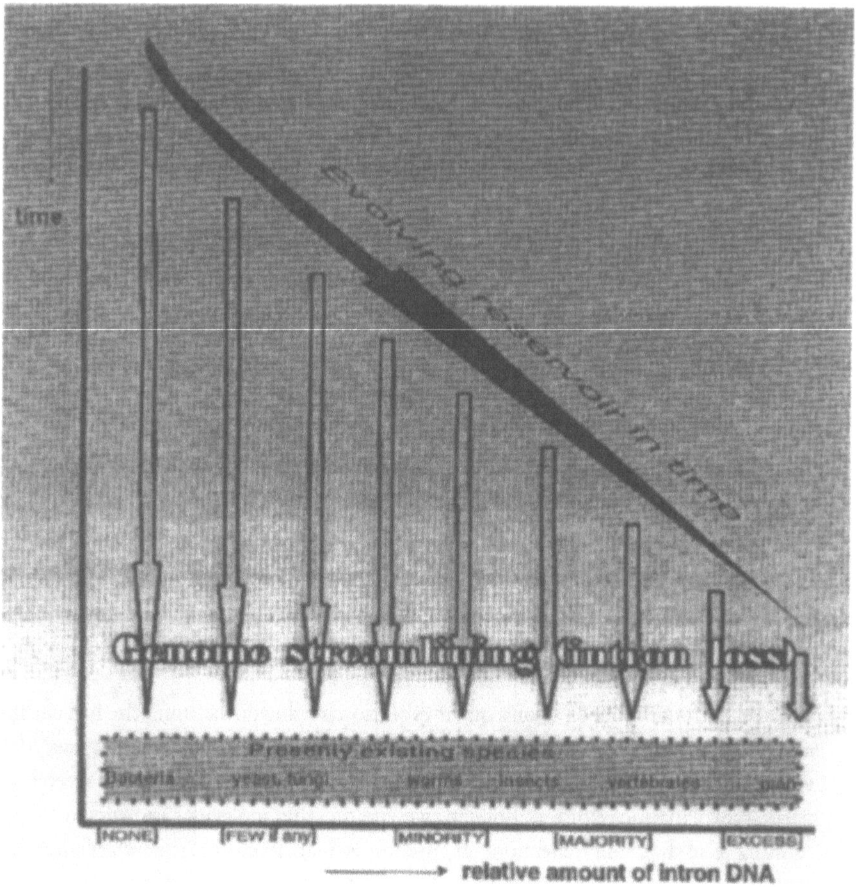

Figure 2. Evolving reservoir in time

phenomena in evolving gene sequences which can be regarded as signatures of retardation processes. By no means should we try to prove or disprove retardation from the molecular studies only; here exactly the contradiction would arise, because retardation phenomena cannot be properly studied and understood without the context of morphology. This workshop was set up with the intention (hope?) to develop a view on genome evolution which is compatible with morphogenesis, in a manner different from what is usually presented in classical molecular genetics or (controversial) alternative field theories.

Literature

Blake, C.C.F. (1978) Do genes-in-pieces imply proteins in pieces? *Nature* 273, p. 267

Cavalier-Smith, T. (1991) Intron philogeny: a new hypothesis. *Trends Genet* 7, pp. 145-148

Darnell Jr, J.E. (1978) Implication of RNA-RNA splicing for eucaryotic evolution. *Science* 202, pp. 1257-1260

Doolittle, W.F., Stoltzfus, A. (1993) Genes-in-pieces revisited. *Nature* 361, pp. 403-404

Doolittle, R.F. (1995) The multiplicity of domains in proteins. *Ann Rev Biochem* 64, pp. 287-314.

Doolittle, W.F. (1978) Gene-in-pieces: were they ever together? *Nature* 272, pp. 581-582

Forsdyke, D.R. (1996) Different biological species broadcast their DNA's at different G+C wavelengths. *Jrl Theor Biol* 178, pp. 405-417

Gilbert, W. (1978) Why genes in pieces? *Nature* 271, p. 501

Gilbert, W. (1987) The exon theory of genes. *Cold Spring Harb Symp Quant Biol.* 52, pp. 901-905

Gilbert, W., Glynias, M. (1993) On the ancient nature of introns. *Gene* 135, pp. 137-144

Goodman, M. (1985) Rates of molecular evolution: the hominoid slowdown. *BioEssays* 3, pp. 9-14

Goodman, M. (1996) Epilogue: a personal account of the origins of a new paradigm. *Molec Phylogen Evol* 5, pp. 269-285

Goodman, M., Tagle, D.A., Fitch, D.H., Bailey, W., Czelusniak, J., Koop, B.F., Benson, P., and Sligtom, J.L. (1990) Primate evolution at the DNA level and a classification of hominoids. *Jrl Molec Evol* 30, pp. 260-266

Kersanach, R., Brinkmann, H., Liaud, M.-F., Zhang, D.X., Martin, W., Cerff, R. (1994) Five identical intron positions in ancient duplicated genes of eubacterial origin. *Nature* 367, pp. 387-389

Kwiatowski, J., Krawczyk, M., Kornacki, M., Baily, K., Ayala, F. (1995) Evidence against the exon theory of genes derived from the triose-phosphate isomerase gene. *Proc Nat Acad Sci USA* 92, pp. 8503-8506

Leipe, D.D. (1996) Biodiversity, genomes and DNA sequence databases. *Curr Opin Genet & Devel* 6, pp. 686-691

Li, W.H., and Tanimura, M. (1987) The molecular clock runs more slowly in man than in apes and monkeys. *Nature* 326, pp. 93-96

Logsdon, J.M., Tyshenko, M.G., Dixon, C., Jafari, J.D., Walker, V.K., Palmer, J.D. (1995) Seven newly discovered intron positions in the triosephosphate isomerase gene: evidence for the introns-late theory. *Proc Nat Acad Sci USA* 92, pp. 8507-8511

Long, M., DeSouza, S.J., Rosenberg, C., Gilbert, W. (1996) Exon shuffling and the origin of the mitochondrial targeting function in plant cytochrome c1 precursor. *Proc. Nat. Acad. Sci. USA* 93, pp. 7727-7731

Long, M., DeSouza, S.J., Gilbert, W. (1995) Evolution of the intron-exon structure of eucaryotic genes. *Curr Opin Genet & Devel* 5, pp. 774-778

Mattick, J.S. (1994) Introns evolution and function. *Curr Opin genet & Devel* 4, pp. 823-831

Patthy, L. (1991) Exons – original building blocks of proteins. *Bioessays* 13, pp. 187-192

Sidow, A. (1996) Gene duplications in the early evolution of vertebrates. *Curr Opin Genet & Devel* 6, pp. 715-722

Stoltzfus, A., Spencer, D.F., Zucker, M., Logsdon, J.M., Doolittle, W.F. (1994) Testing the exon theory of genes; the evidence from protein structure. *Science* 265, pp. 202-207

Verhulst, J. (1993a) Louis Bolk revisited: I. Is the human lung a retarded organ? *Med Hypotheses* 40, pp. 311-320

Verhulst, J. (1993b) Louis Bolk revisited II. Retardation, hypermorphosis and body proportions of humans. *Med Hypotheses* 41, pp. 100-114

Verhulst, J. (1994) Speech and the retardation of the human mandible: a Bolkian view. *Jrl Soc Evol Syst* 17, pp. 307-337

Verhulst, J. (1996) Atavisms in Homo sapiens: a Bolkian heterodoxy revisited. *Acta Biotheor* 44, pp. 59-73

Verhulst, J. (1997) Der Erstgeborene. Mensch und höhere Tiere in der Evolution. Verlag Freies Geistesleben, Stuttgart, *in press*

22 Genetic disabilities – predictive diagnosis, gene therapy and communal care

WORKSHOP

PETER MINY
Clinical geneticist and genetic counsellor
University of Basel
Römergasse 8, CH-4005 Basel

NICK BLITZ
GP and medical adviser
Camphill Medical Practice
Murtle, Bielside, Aberdeen, AB1 9EN, UK

PETER MIDDLETON
Physician, pulmonologist
Westmead Hospital
AUS-2145 NSW Westmead

Reporter: Pat Cheney (UK – Gwynedd)

Abstract
An examination of predictive genetic testing, its usefulness in practice and its implications for individuals and society, was followed by an account of the first European gene therapy trials for people with cystic fibrosis. Finally our definition of disability was questioned in the light of clinical and social experience.

Presentation by Peter Miny
Peter Miny gave an account of genetic testing, restricting himself to predictive testing, a simple definition of which would be 'pre-symptomatic' testing. These tests are done on healthy individuals to determine the presence of mutations which might lead to clinical disease, where there is a known increased risk for a genetic disorder. Where the tests show an exclusion of the mutation then there is almost no risk of the genetic disorder;

where they confirm the mutation then there is a risk. Several arguments for and against genetic testing were put forward. For example, relief which arises after the exclusion of a specific mutation, a possible cure by early treatment or individual life and career planning in the case of no cure, were seen as advantages. By contrast, there is the psychological burden of receiving a positive test result, especially where the therapeutic consequences may be unclear. At the present time there is also a lack of definite regulation to keep genetic information confidential with regard to insurers, banks, employers and others.

In Peter Miny's view, many people may benefit from predictive testing. Everyone should have the right to choose to have a test but no-one should be compelled to do so. Furthermore, there should be a strictly individual approach involving proper genetic counselling. A very good paradigm for this is the one used for Huntingdon's disease. Confidentiality is of critical importance in any genetic testing.

Prenatal predictive testing

This has been done for twenty five years for congenital genetic diseases such as chromosomal abnormalities, metabolic disorders and Duchenne muscular dystrophy for example. Technically speaking, all post natal testing can be done prenatally though this gives rise to serious difficulties and questions and adds the ethical dilemma of elective determination, especially in late manifesting diseases (in the case of a predisposition to breast cancer, Huntingdon's Disease for example). The aim of such testing would be early therapy.

The two prenatal predictive tests used are amniocentesis and chorionic villus sampling. The indications for a placental biopsy would be suspicious ultrasound findings. The possibility of chromosomal abnormality will increase with maternal age. The 1994 statistical yearbook says that out of the 75,000 terminations of pregnancy, 1.12% were for genetic reasons, though not all terminations will find their way into this figure.

Pre-implantation and pre-conception genetic testing

In Peter Miny's view and that of many others in the field, this form of testing is considered to be a superior option for replacing the traditional prenatal diagnosis. It avoids termination of pregnancy but it is restricted to high risk situations since In Vitro Fertilisation and embryo transfer is required.

Four examples of specific conditions and their consequences were given, in a simple form.

Huntington's Disease In this first example, we have an illness which begins with a movement disorder and ends with mental handicap. There is a decreasing risk of manifestation with increasing age where no symptoms appear. It would be simple to screen everyone, however, there is no therapy at the moment. Given this situation, one is compelled to ask how the genetic test got into clinical practice. In terms of patient organisations those for and against testing are split 50/50 and in practice fewer than 50% of the people who are at risk take the test. In Holland 15% of the population at risk is tested.

Polycystic Kidney Disease The second example is of multiple cysts on the kidneys, necessitating renal transplant. Knowing the details does not change the therapeutic approach but genetic testing may replace invasive testing.

Adenomatous polyposis The third example is that of intestinal polyps. People with this disease undergo microscopy at intervals. A negative test would mean no regular clinical workout. Prophylactic treatment (surgery to remove the colon) is given to affected individuals. This is an example of early diagnosis being positive for the patient. It is also an exception to the general rule that genetic testing should not be done on minors.

Breast Cancer The fourth example is that of breast cancer. The risk factors are viewed in relation to the following factors:

- age
- gynaecology and pregnancy
- exogenous factors (e.g. hormone)
- family history
- predisposing mutations

The percentage of risk associated with family history is a continuum, from no cancer at all in the family, in which case the risk is assessed at 0-1.3%, to two or more relatives with cancer, where the risk rises to 91%. Ethnic grouping is also a factor in predisposition.

It is not clear which therapy is appropriate for breast cancer in the presence of any particular mutation and it is therefore at the moment only useful for prediction. Excluding the manifestation does not mean that breast cancer will not arise but only increases the risk. So predictive testing is very simple in the case of Huntingdon's Disease and quite complicated in breast cancer.

Discussion

The discussion began with a question about whether there are any figures for the numbers of false positives and false negatives in the testing procedures. According to the presenter, if chromosome analysis is done properly the risk is extremely low. Quality assurance is improving all the time but money is always a problem. In eight years of chorionic villus testing, Peter Miny knew of only three Down syndrome children who had been born in spite of the test. This is seen as a low figure in the light of hundreds of thousands of tests. There is an overall risk of 1-2% abnormality in newborns outside the testing.

The relative merits of testing/non-testing and the question of choice were then addressed. It was felt that there is huge difference in for example, prenatal testing for

Down syndrome and testing for a predisposition to cancer. What is the value of diagnosis in general where there is no possibility of therapy? In Huntingdon's Disease for example, some people want to know whether they have the possibility of the disease for the purposes of deciding whether or not to have children, though interestingly, not all of those diagnosed positive decide *not* to have a family. Peter Miny stressed that everyone who has a test is counselled. In general there is a lack of financial resources for counselling although in the United States some programmes have justified themselves on the basis of reduction of genetic disease. This is by no means the case everywhere. When it comes to insurance, some companies will pay out if parents promise to abort the foetus.

The fundamental dilemma of individual choice in the context of a genetic screening programme was raised. If there is a theory of genetic disease (Down syndrome being one example) and a screening programme is introduced on that basis, where does freedom of choice arise? The language we use is full of metaphors which come out of genetic thinking. Does the symbolic power of prenatal diagnosis lie in the fact that we talk about choosing 'healthy' children? Whereas it is generally accepted that counselling should be as far as possible 'unbiased,' in the Netherlands for example, it is obvious that the counselling is not non-directive. The breast cancer screening programme suggests a possibility of control and management which is very misleading. Is not genetics too much the expression of a dangerous and scientifically inadequate model for humanity?

In prenatal diagnosis should we look at the viewpoint of the child or the parent? The result of a diagnostic test for Down syndrome is apparently looked at from the parents' point of view but is it not possible to consider testing from the unborn child's viewpoint? How could we imagine what the child's point of view might be? In the United States a Down syndrome child brought a legal case against his parents for not having him aborted, and won. Faced with such imponderable questions, where we feel unable to bring any therapeutic benefit to disabled people in our society, perhaps it would be wiser to simply accept them.

Money is a limiting factor on which diagnostic treatments are being offered. At the present time we have a situation of society versus science, whereas society could be seen as a mirror of science. The government could give shape to the social response. Science, which gives the illusion of dealing freely with health and disease, is a compulsion, whereas it should be enlightenment. It should not be repressed, in societies where art and science are oppressed there is great horror, but we should see 'real' scientists as artists transforming matter and the other scientists as journalists discussing and commenting on it. Scientists could come to people with what they had to offer. For example, people who had been told that they had the possibility of a Down syndrome child could meet people with Down syndrome and get to know them.

Presentation by Peter Middleton

Peter Middleton worked with Duncan Geddes and Eric Alton in London on the first European gene therapy trial in human subjects with Cystic fibrosis. Initially, he outlined the history of our understanding of the disease, beginning with a quotation translated from German folklore:

'woe to the child that tastes salty when he is kissed for he is bewitched and soon must die.'

This quotation dates from around 1700 but it wasn't until 1938 that the clinical syndrome was recognised. In 1953, note of the increase in sweat electrolytes and excessive salt loss was made by Paul di Sant'Agnese. In 1981 the nasal potential difference and its relation to sodium absorption was documented by Mike Knowles and in 1983 the defect in chloride transport in the sweat duct was noted by Paul Quentin.

As far as we know, essentially all people with mutations in their CF genes develop the physiological abnormalities of CF, together with the pathological changes found in the disease. This is an example which illustrates the reductionist principle that a specific defect gives rise to a specific disease.

CF was localised to a mutation on Chromosome 7q in 1985. In 1989 the gene was isolated and termed the cystic fibrosis transmembrane conductance regulator (CFTR). Cell lines were 'transfected' with this gene in 1990, and knockout mouse models developed over the next 2-3 years. These mice have only some of the problems of cystic fibrosis found in humans, which does not support the simple reductionist view.

Genotyping of patients showed that up to 1 in 20, to 1 in 25 of the Caucasian population will have a mutation in the CF gene, which represents the CF carrier risk. The most common mutation is the Delta F508 mutation which accounts for approximately 70% of the mutations in the European population overall. Different ethnic groups (i.e. Ashkenazi Jews) will have a different spectrum of mutations. There are different classes of CFTR mutations causing CF and different mechanisms that may cause the disease. There is a continuum, from the presence of the mutation in infertile men who do not manifest any other CF symptoms, to those who manifest them all.

Gene therapy is based on the idea that we transfer new genetic material to the cells of an individual with resulting therapeutic benefit to that individual. In the case of CF, the gene was just added to the nose with the aim for it to function locally. The strategy involves the somatic cell line, (those of the body itself), with no effect on the germ cells (sperm & ova). The transfer system can be a liposome – DNA plasmid conjugate or a recombinant virus vector. The aim is the reversal of the cell defect; prevention of lung damage; lessening of recurrent infections and progressive respiratory decline together with other effects on the pancreas and liver. Of the two delivery systems, the adenovirus vector system and the plasmid based system – the latter is considered safer. The therapy was offered to the CF mice via a nebuliser and there was a change in electrophysiology following gene delivery.

At the Brompton Hospital, London, the gene therapy trials were randomised, double blind and placebo controlled. The gene liposome complex was delivered as a single application with a nasal aerosol in three different doses: 10, 100 or 300 micrograms per nostril. The fifteen patients were all male, homozygous for the Delta F508 and with no history of nasal operations. Nine were given the active treatment and six the placebo. Measurements were taken of the nasal PD and a biopsy was performed. The results showed small but encouraging physiological changes and the conclusions were that liposome mediated transfer is safe and can partially correct CF ion transport.

No gene therapy should be undertaken without strict safety measures. In the United States, where a gene therapy trial in human subjects was also being undertaken, the method of gene transfer used was the adenovirus vector. This resulted in inflammation in one of the patients requiring admission to intensive care. Neonates have not yet been studied, though in the future, treatment would ideally be started as soon as diagnosis has been made, before disease has caused any permanent damage.

For many genetic defects, it can be argued that there is little direct correlation between cause and effect i.e. gene equals illness, but in CF this has been the case. Even though CF is a classical genetic disorder, we are not yet at the therapy stage as there is so much more to know about gene therapy. One of the main difficulties with all of these techniques is that we do not know how much gene in how many cells is required to prevent or treat the disease. More will need to be known about cystic fibrosis itself. For example, there is considerable variation in the way cystic fibrosis affects different individuals with the same genetic mutation, suggesting that other genes or environmental factors play a role in the determination of the phenotype.

Overall, it is unlikely that gene therapy will be standard medical care for CF for at least five years.

Discussion

Following on from Peter Middleton's final remarks, one of the doctors present said that he had used homeopathic medicine on the liver, pancreas and intestine of CF patients and seen a favourable effect in much less than five years.

It was asked whether gene therapy only has a chance where there are localised problems. The answer was a qualified, 'yes, at the present time'. The therapy has been used on blood vessels in the leg for instance and where the affected tissue is accessible, as in the lungs of CF patients. Publicly, there is a great deal of unrealistic expectation for which the press is felt to be largely responsible, but really not much progress. However, the therapy is not far from being of clinical use in certain specific cases (lung cancer for example). Gene therapy may be of more use in acquired diseases such as lung cancer than in genetic diseases.

A question arose about the advisability of allowing the extraction of sperm from CF men. Is this ethically justifiable and should we consider its effect? The side effects of the treatment are certainly a problem: i.e. the general anaesthetic for the wife for the removal of ova and the actual extraction of the sperm from the husband. The wife would need to be checked for the defective gene of course. Furthermore this method has a low success rate in CF men and there is still a small risk of a CF child even if the mother is negative for Delta F508. The discussion then moved to the high prevalence rate of CF mutations in the general population. There are many different hypotheses to try to explain this: fertility in CF heterozygotes may be greater than that of the general population or possibly the mutation gives some protection against cholera or *E.coli* bowel infections. Permanent changes in chloride channels can be helpful for instance in cases of botulism.

Presentation by Nick Blitz

Nick Blitz began his presentation by suggesting we look at certain assumptions in relation to normal/abnormal, different, handicapped, disabled. He referred to the example already quoted in the conference, of the broken clock as a metaphor for the human being. The clock may not be working but it is nevertheless still a complete clock even if it no longer serves the function of timekeeping.

'When it comes to defining disability, we are all disabled to some extent, either cognitively, socially or physically. We tend to assume that disabled means dependent. Is this true? In fact we are all interdependent. The assumption behind this idea is the inability of disabled people to contribute to society as we see it. We assume that disabled people are less useful and because they don't use the things in society that we assume make a good life, we judge their quality of life to be poor or inferior. The medical model presupposes abortion; the social model talks about integration and normalisation. So disabled people get a mixed message from society, a double standard which is confusing.

In the context of care the beliefs of the carers are important. The genetic idea is that by chance things have gone wrong. When not aborted then care is provided, *but* the holistic view posits soul and spiritual integration. One can say that spiritually a disabled person can be viewed as intact and that this is the case with every person, all of whom are disabled in some way. Working in this way provides a picture of meaning and the possibility of a shared need for growth and change. In other words it is not just a question of educating the disabled person but of recognising the need for a mutual change and development (carers and cared for), that is the context of curative education and social therapy.'

A video was then shown, which everyone was invited to look at from two points of view. First, in considering the relation between phenotype and genotype, is there a direct causal relationship? Secondly, what can we learn here about polarity and archetypes?

The video began with David, a fifteen year old, sitting with an adult in a garden. He looked young for his age. He did not speak (it was later pointed out that he has no language ability), but sat with a fixed grin on his rather angular face. His movements were ataxic and his hands and feet rather large. He apparently has a poor appetite – he is extremely fussy about food and has a very low sleep requirement. His grin is an expression of his constant anxiety, exacerbated by his poor comprehension. He also suffers from epilepsy. He has what is known as Angelman's syndrome.

The second extract was of Gerard, who was nineteen but looked older. He is of a strikingly different physical constitution to David. He is short and obese, round and soft, with a lack of muscle tone. He has small hands and feet and suffers from hypogonadism. His obesity is the result of a preoccupation with food – hyperphagia – and in relation to this he is very manipulative. He also sleeps a great deal, even in the daytime. Communication is easy – he loves to talk and will always engage someone in conversation if they are near him. Emotionally he is very unstable. All this is fairly typical of the disorder. Gerard has Prader Willi syndrome.

The polarity of Gerard and David is interesting. Aspects of it manifest a certain male/female polarity. Genetically they both have a disorder of Chromosome 15 – the same locus in each one. There are a number of genes on that locus. Prader Willi syndrome involves a micro deletion of paternal origin and Angelman's lacks the maternal contribution. In view of this, can one broaden one's view of the relationship between the gene and its manifestation?

The third film was of Alan and Hannah, two adolescents with Down syndrome, sitting and conversing easily in a garden. They were obviously close friends, speaking with ease and confidence and apparently feeling at home in the world.

How do we understand Trisomy 21 and its manifestations? In his paper 'The Environmental Basis of the Down Syndrome Phenotype' Burton Shapiro (1994) says that morphogenesis and development take place within an epigenetic landscape. The model is of a downhill slope with grooves in it that interweave and overlap. Our development, in simplistic terms, proceeds down a groove that approximately meets the normal developmental pattern. Depending on environmental stresses, and these can include genetic stresses as in Down syndrome, the homeostatic mechanisms that keep us on the more or less normal groove will be disrupted or not. There is not a single feature in Down syndrome that is not found in the general population. In other words the features are not specific for Down syndrome. With various autosomal Trisomys many share the same physical manifestations. There is a lack of concordance in identical twins with Trisomys. There are stable and less stable factors in morphological development and it is the less stable which are affected by the Trisomy (i.e. the development of the simian crease in the embryonic hand which is a feature of Down syndrome). These are thought provoking in terms of the genetic model and also in terms of prenatal diagnosis predictions. The spectrum in Down syndrome is enormous and one cannot tell by prenatal diagnosis how severely affected the individual will be.

As a final remark Nick Blitz mentioned the statistical relationship between the mother's age (ova theory) and the incidence of Down syndrome. We could look at the very particular relationship that exists between Down syndrome children and their mothers. It is very close. Maybe it's not just chance but has an existential meaning.

Discussion

The discussion centred around the appropriateness of the 'DNA thinking' approach in looking at disabilities. Is it too circumscribed? Is gene therapy a promising avenue to pursue at all or are the concepts behind it inadequate? For that matter, is medical treatment of any kind appropriate?

When we talk of the environment, perhaps we should be careful to define the limits of that environment and even take into account social and spiritual aspects. It is possible that the environment determines the disability and also the individual themselves. Although we do not know the answers to these questions it is nevertheless possible to use something without knowing how it works. Every new form of medicine is given before we know how it works. When, for example, Alexander Fleming discovered that penicillin could be used to fight infection, he didn't know how it worked. This is analogous to gene therapy: there is a limited amount of knowledge and all the steps between A and B are not known.

Reference

Shapiro, B. (1994) The Environmental Basis of the Down Syndrome Phenotype, *Developmental Medicine and Child Neurology* 36, pp. 84-90

23 Heredity, gene therapy and religion

WORKSHOP

ULRICH EIBACH
Theologian
Ev. Theologische Fakultät, University Hospital of Bonn
Sigmund Freud Strasse 25, D-53127 Bonn

MARTIN GMEINDL
Gynaecologist, obstetrician
Department of Gynaecology, Gemeinschaftskrankenhaus Herdecke
Beckweg 4, D-58313 Herdecke

Reporters: Manfred Schleyer (D – Neuhausen) and David Heaf (UK – Gwynedd)

Introductions

Ulrich Eibach launched the workshop on the first day, which focused on prenatal diagnosis and ethical problems in genetic testing and counselling. In no other area of medicine is the difference between what can be diagnosed and what can be treated as great as it is in prenatal genetic diagnosis. It gives rise to a new situation in medical practice: that of a scientific diagnosis without any therapy existing to cure the disease. Usually diagnosis of disease is ethically justified only if it is in the interests of the patient. This gives rise to the question: to what extent can we justify such diagnoses knowing that there is no therapy? The Catholic church for example takes the view that this is unacceptable. One consequence of such diagnoses, without the possibility of therapy, is the prevention of the birth of handicapped children by abortion. The diagnostic test results therefore necessitate a decision about the value of life and raise the question whether diseases and handicaps make life worthless. Who should be the judge of this and should such judgments be widened to embrace those who are already born and those who are still descending to birth.

It has been asserted that diagnosis without the possibility of therapy leads to euthanasia. To what extent is this idea challenged by the concept of the 'object of therapy', namely the idea that the diagnosis is driven by the question of the family's or society's ability to carry the burden of a handicapped child? Does this mean that in practice no fundamental question about the worth of life is actually addressed? Is the right to reproduce a fundamental right even amongst people who carry the genetic risk of producing a handicapped child?

The concept of the 'object of therapy' gives rise to even more questions. For instance could society itself be an 'object of therapy' and as such could the way in which handicapped people are accepted by society be a criterion of it? If so, who decides whether society can accept these people?

Genetic analysis enables the diagnosis of diseases and handicaps that manifest before, during and after birth as well as others which may arise later in life or which are enhanced by genetic factors. The spectrum of diagnosable diseases constantly increases. It ranges from severe monogenetic and polygenetic diseases, to less severe diseases and handicaps and to genetically enhanced diseases such as psychoses and epilepsy. It is important to differentiate the monogenetic from the polygenetic disorders in any plans for therapy. Everybody wants their child to be healthy and society as a whole wants to promote the birth of healthy children as much as possible. We are again faced with the question: who decides what is acceptable? Should selling genetic tests for diseases be permitted and if so, how and by whom should it be regulated?

The German association *Medizinische Genetik* (Medical Genetics) requires that genetic testing takes place together with genetic counselling. At present this is the case in about a quarter of diagnoses. Genetic testing is done as a matter of course and physicians have to offer it to all risk groups in order to avoid the possibility of having to pay compensation for the birth of a handicapped child. Genetic counselling aims to enable affected people to make an informed and independent personal decision. But what criteria help them to decide? And is there any point in having a genetic counsellor if he has no say in the individual's ultimate decision?

The predominant view in society is that it is irresponsible to reject the opportunities of prenatal diagnosis and consciously accept the risk of bearing a handicapped child because the community has to pay for the consequences. It is seen as important to prevent the birth of handicapped human beings. The result of these social constraints is that handicapped people experience a threat to their personalities and their rights. Genetic testing leads to discrimination against whole groups of people. They face a denial of their rights to reproduce or just to exist. Also, the value of researching into genetic causes of mental diseases is questionable. Scientific paradigms influence the attitude of society to handicapped people and the incurably ill.

Should the handicapped themselves and their representatives decide on whether genetic tests should be introduced? People argue that anyone faced with genetic testing is under no compulsion to make use of it and that therefore genetic testing should be available to all.

Martin Gmeindl introduced the workshop on the second day. Modern medicine is confronted with two nearly unsolvable ethical problems: on the one hand the transplanting of organs based on a sensory view of death and on the other hand, birth threatened by manipulative operations, prenatal testing, in vitro fertilisation and genetic manipulation. When it comes to ethical problems, researchers in science seem to be at a loss and appear to exclude such problems from their work. Instead they are dealt with in ethical committees which look to the arts and theology for help. Yet the ethical consequences are severe. We should keep in mind that National Socialism was partly based on a reductionistic view in research on heredity. This image of the human being soon reveals its inherent contempt for mankind. As a result of this the ethical problems are especially controversial and require special vigilance.

True gene therapy, namely the manipulation of the genome of cells, is only possible if a variety of research results are used, a few of which include manipulation of bacteria, plants and animals, in vitro fertilisation, embryo implantation, prenatal diagnostics and genome analysis. Many of these methods are misused, for example, for the determination of a child's sex and, by selective abortion, for denying them life. Whether gene therapy can be used for the welfare of mankind remains to be seen. Up to the present time, no true gene therapy experiment has been more than a clinical trial. As a society we are experiencing that many scientific insights realised through gene therapy trials can be grossly misused. As to the associated ethical problems, we see a severe worldwide powerlessness and a subtle anxiety as to whether insights into birth and death arising from these experiments are being used to the benefit of mankind. The hope is that it will be possible to control and cure severe diseases by gene therapy involving the skill and ingenuity of engineering on a molecular level. A start has been made on some inherited diseases and cancer. The implication of this is that any readiness to consider the possible meaningfulness of certain diseases is rather low. All hopes are concentrated on a stroke of biochemical genius. The research aims mainly to control diseases and not to understand them at all. Thus it remains to be seen if interest in the diseases decreases if successful therapies become available and it must be asked given this situation whether a deeper consideration of the nature of certain diseases and overcoming them remains possible in medicine. True therapy arises from the co-operation between therapist and patient. Through such co-operation a problem that exists can be solved in the best way possible. It involves the therapist's selfless engagement on behalf of the patient. In so doing, both the patient and the therapist gain new strengths.

The ethical problems we are faced with through the development of gene manipulation and its consequences are not solvable by science. We cannot find the strength to overcome this crisis by science alone. Here we need the kind of help we will only receive if we see and accept our powerlessness. In this way we will increase our ability to achieve a true understanding of the nature of mankind. Only then can we understand and circumvent developments which harbour cruelties. Realising that the knowledge gained by natural science is inadequate for dealing with the problems I have outlined is a significant threshold. It is important to question what needs arise from these

experiences. Without the benefit of this man becomes entangled in developments which might overwhelm him. Many people contributed to what happened fifty years ago. But no-one on reflection would have wanted it.
It was obvious from the beginning of the discussion which followed Martin Gmeindl's introduction that the complexity of the topic made it impossible to give simple answers to the questions. We agreed that each individual situation has to be examined carefully. The discussion broadly centred on six distinct topics according to the paragraph headings below.

Science and scientists
What motivates research? Scientists begin research mainly out of curiosity, wanting to know. Only a few scientists are motivated initially by the desire to help. In doing research, questions of career, success and economic concerns arise. In the USA in particular research funding is aimed at obtaining something usable. Company financial interests and the belief that society must be manageable by health administrators are involved too. Scientists' thinking splits between the work and its application. This leads to a science without ethics. In the view of scientists, ethics has to be determined outside the workplace by philosophers. If this 'schizophrenia' is recognised, scientists often become ill. Reflections are necessary at some point in research. For example 'what am I doing?' or 'how many mice have I used?' In this respect scientists are responsible for the consequences of what they do. If they discover new medical possibilities, then patients who want to be healthy call, without any lengthy reflection, for diagnostics and therapy. Another factor has to do with tradition as an ethical guideline. If the reality of the spirit or religion is lost, one believes only in what one sees – that is materialism. This culminates in the loss of our Christian culture. As the transmitted and accepted traditions vanish, the loss of ethics in science accelerates. This loss reached a peak during the time of National Socialism and today too materialism encourages a creeping loss of ethics. In Germany in particular at this time people fear the research. Science has no answers. We as a society have to give the answers.

The way in which disease is viewed
We experience that our view of diseases directs our feelings and actions. Here too science shapes our thinking and emotions – sometimes in an inadequate way. In general we want to eliminate all disease. We want to be healthy and have healthy children. It is possible to say that in general a realistic view of illness does not exist. Take for example, Down's syndrome which can now be diagnosed prenatally. People think such children are, to use an extreme expression, monsters. Such people are not familiar with the reality of the situation. Often a contact is useful, especially for physicians, to enable them to develop a grounded perception. Then the physician can give a true view of the illness. Otherwise it leads to the 'Singer argument': "Better children without Down's syndrome

than with", even though most people have no experience of handicapped children and only know about these things from descriptions. Such children are happy and in general are like a kind of sunshine in a family. We learn from growing children and experiences like this enable family members to become socially competent.
Another example is Alzheimer's disease. This can manifest from as early as thirty to forty years of age and shows an increasing loss of awareness in those affected. The patients become confused. With the introduction of a genetic test for Alzheimer's disease there is now a possibility of abortion where the diagnosis is made. But with abortion thirty to forty years of life are extinguished. Who will decide that these years are useless or senseless? What about thirty happy years? Would these be worthless years? Before living life really consciously, some of us may have to wake up to the fact that each day can be a gift! If we want to develop a new kind of thinking about disease, we must raise these subtle questions again and again.
Yet another example is the current research into epilepsy. The picture of epilepsy is connected with a functional interference of brain processes. Where epilepsy occurs specific aberrations in an EEG can be shown. But in only 5% of patients can these aberrations in the EEG be linked directly to the illness. Today there is a strong urge to discover a genetic basis for epilepsy. If it is found with this EEG phenotype, the result will be the abortion of those affected, 95% of whom would be healthy. Yet parental fears will be decisive.
The same holds true for muscular dystrophy, whose sufferers die at about age 20 – 25. It can be diagnosed with a probability of 25%. Who can cope with the problems that arise if the parents decide that they do not want to abort but want to have the child. No-one has to accept a gene test, but we have to fight against 'good advice' and parents have to cope with the enormous suffering if a handicapped child is born.
This leads to the general question: what is disease? Only 2% of known diseases are hereditary. In all other cases other influences cause or are linked to the specific disease. The struggle against disease takes precedence over a quite different kind of reality. Fear of sickness preoccupies our thinking. But looked at in a different way, we might consider whether disease is meaningful. In so doing we can see that the experience of disease often changes our perception of the world. For the most part we heal ourselves. Only occasionally is it the therapy, and it is possible in special cases for a physician to heal without any accompanying therapy. Disease is meaningful. We as individuals are more important than our physical body and we can learn a lot about ourselves through disease. It may even happen that we have to change our lives in order to remain healthy. If we see disease as simply a troublesome event, we are thinking mechanically. An attitude towards the doctor develops which is one of: come – operate – leave. We miss the possibility of seeing the disease as a starting point for therapy of both the physician and the patient. Diseases can be seen as problems to be solved by ourselves or by others. At the present time we are inside the Cartesian split: my body and me. My body is not

me but nevertheless I must still care for it. Instead of a technical type of therapy we need a comprehensive therapy.

In former times Christian cultures accepted disease. What do we think of diseases now? We find them hard to accept. Jesus healed without asking people whether they wanted to be healed. He healed if one moved towards recognition of him as Christ. The story of ten lepers was used as an example: the only one to be healed was the one who believed in Jesus. The patient has to participate and physicians have to be ready for teamwork in healing and therapy. But people reject doing something for themselves. We have a consumer mentality. Doctors are not recognised as helpers, they are seen as mechanics. Now genes are held to be responsible. If you are depressed, it's the fault of your genes. *You* do not have to change. Ultimately this leads to dangerous thinking. In this respect the deciphering of the human genetic code – the HUGO project – will answer only a limited set of questions. We have to look at causes not at genes. Therapy for alcoholism was raised as an example. Drugs exist which block the receptor for alcohol in the brain, but unfortunately they cure only some of those affected. Regardless of the underlying cause, the therapy depends on the doctor. One can prescribe drugs or look for the underlying problems.

The difference between feelings and thinking in relation to diseases

Depending upon the situation, we experience that a difference exists between our feelings and our thinking concerning diseases. If a child is stillborn, only in a few cases do parents want a funeral. But if the child dies during birth the parents ask for a funeral. Often the difference is only a few days. No funerals are wanted for aborted children. This raises the question of where our view of the human being comes from. Who defines a human being? Certainly not science! It was stated that the value of being human is related to a feeling of inner dignity. If we experience this dignity we do not need a witness or judge. Value does not come from the outside. Confronted with the situations described, one can feel a kind of loneliness and self-doubt. We can find ourselves or be lost.

The value of life

We feel ourselves living in a world where our personal dignity is attacked by the current reduction of all life to just chemically active matter. In former times the value of a man rested on his simply being human. But this view changed in time. Man's value was once seen in relation to God. Then with the advent of Darwin mankind came to be seen as descended from animals. This approach reached its culmination at the time of National Socialism, where only one particular race was said to be truly human. The view that some lives have no value began to run its terrible course. This raises the question of setting limits. The contemporary debate has the potential to become dangerous and can be compared to an apocalyptic development. If we start discussing what is and is not

human we can arrive at the idea that some lives are of no value. When we think of heredity in these terms we build up a certain picture of man and are in danger of asserting our views as reality. Relativising truth is a matter of concern. If truth cannot be found any more we as a society have to draw up general rules and set limits.
In a Christian sense we are wanted by God and at the same time God is the source of our dignity. We can get our intrinsic value in different ways: through God, through being human or just by being alive. But religious views are not accepted in a secular society. The experience of National Socialism and the consciousness of the holocaust which arose from it mean that it is difficult to discuss this topic, especially in Germany. In the Christian tradition these matters are a test for civilisation. How will it respond? If its response is restricted to the point of view of natural science, for example that of Peter Singer, we step out of tradition and civilisation.

The situation of handicapped people

A connected issue is the situation in which handicapped people find themselves: increasingly they are not accepted. We were told of one handicapped person who hears more and more often words such as, 'Nowadays someone like you would not be born.' In hospitals the situation of invalids gets worse if they have no surviving relatives. In such cases the physician has to decide what to do. It was said that in extreme circumstances the patient is left without food until he dies.

Ethical questions

At the present time we are experiencing the loss of religion, of fundamental moral instructions, of God. That is to say, freedom! There is no longer a moral consensus. We have to be aware ourselves of what we are doing. We should not be forced to do things by friends, relatives or patients.
The problem of genetic tests was raised in relation to insurance. There are insurance policies which call for healthy babies. But we need a willingness to accept handicapped people. In the USA, people who are not genetically tested risk not being given insurance. However, discussion of this is still in progress. In most cases the inherited diseases are already manifest in the family so that the risk is known before genetic tests are made.
But genetic tests can also provide an opportunity. Results have to be waited for and the experience of waiting changes people. Although in the meantime it is possible to become more informed and question ones situation, it is hard nevertheless to know what to do with such information. In addition it can be very unfair and dangerous to confront people with possibly fatal outcomes before they are ready. This is especially the case if nothing can be done. At all events, such new awareness can be a chance to restructure one's life.
The question of the utility of tests during pregnancy was raised. People do not necessarily want information, they want a happy healthy child. As things stand at the

present time patients have to be informed of any consequences before a test, but only 15% get this counselling. A positive diagnosis is shocking for the parents. Usually it will be the first time that they have had to consider such a thing. They then have to make a decision in a very short space of time, often in less than a week and frequently they decide to abort out of helplessness and fear. In these cases a more balanced judgement is necessary. Each situation is different and there can be no short cuts to a decision. In addition, there is the pressure to abort the handicapped which comes from people's environment: economic pressure, the threat of too much suffering etc. Again the question of economics is part of the picture. There will be economic problems arising from the care of old people in the future. At the present time, 80% of elderly people are cared for at home and in 90% of those instances, by women. This will change in time. Experience tells us that these 'technical' reasons for ethical decisions are on the increase.

In the course of the discussion we became aware that there is much foolish thinking about these matters. We simply cannot refuse humane solutions and treatments for economic reasons. We have to be wide awake to such foolishness, because so many people take these thoughts for the truth. It is we who must decide which arguments are foolish and which are dangerous. For example, at the present time, no one would seriously consider aborting those with a susceptibility to Alzheimer's disease in later life. But what we think now, we will do tomorrow! We have to fight the very beginning of this kind of thinking.

There are developments which are already leading to processes which are very questionable from an ethical point of view. One example is in vitro fertilisation as a matter of course for infertility. It is a quick, and for some couples the only method of getting children. As the success rate is low, physicians fertilise six embryos but (in Germany at least) the implantation of only three is allowed. If more than one or two embryos successfully attach they have to be 'reduced'. But what happens to the remaining cryopreserved embryos? Recently in the UK they had to be destroyed. All these problems are viewed as technical problems. There is no clear solution other than by educating couples to recognise what the problems are. Some couples will not wish to take account of these at all. They hold to the view that one has to have children. This leads to situations such as the one in China where with the help of doctors, parents are choosing to have no girls or only children who will supposedly be intelligent.

It seems that at the present time there are two fundamental attitudes: a struggle on the one hand for a clear awareness of humanity and dignity and on the other hand the crude thinking of 'I do what I want.' But who should set limits? Who decides? Scientists? Politicians? These issues force us to open a general public debate about the basis of our present day morality. As a society we can ask scientists to contribute to the debate but ultimately society has to decide. One urgent need is the education of young people. If things develop in such a way that we become aware that we are unable to find answers and are helpless, it could lead us once again to religion. Increasingly people experience spirituality without calling it God. If we unite our efforts and do something it may help.

24 Genetic engineering and xenotransplantation

WORKSHOP

JÖRG JUNGERMANN
Physician
Gemeinschaftskrankenhaus Herdecke
Beckweg 4, D-58313 Herdecke

PETER BRAIDLEY
Surgeon
Transplant Unit, Papworth Hospital
CB3 8RE Papworth, Everard, Cambridge, UK

Reported by Haijo Knijpenga (CH – Dornach)

After the participants had introduced themselves and shared their questions and motives for attending, Jörg Jungermann introduced the topic of the workshop. Participants' questions referred mainly to the social implications of xenotransplantation, the problems of using animals as experimental objects and to the man-animal relationship in connection with the questions: What is an animal? What is a man?

Jörg Jungermann gave an overview of the historical development of the biological sciences and referred to some essential steps which are preconditions for modern xenotransplantation medicine:

1 Since the discovery of the system of the ABO blood groups, blood transfusion has become an established replacement therapy. On a cellular basis, it is a transplantation from person to person.

2 Before 1962, transplantations were unsuccessful because of the immune barrier. Thereafter the development of immunosuppression (azathioprine-corticosteroid) made them practicable, but their success was limited by further immunological problems. In 1963 a patient survived 9 months after receiving a chimpanzee kidney transplant. Other xenotransplants with baboon heart (1984) and liver (1993) failed. Homologous

transplants (kidney, liver, pancreas) were more successful. But attention was paid to these mostly by the scientific community and hardly at all by the general public or general practitioners.

3 In 1967 a shock wave went around the world following the heart transplant performed by Dr Christian Barnard. This event was followed by a strong division of opinions into pro and contra. In the meantime cyclosporin, an immune suppressor had been developed.

4 In 1968 came the far-reaching decision by the Harvard ad hoc committee. It redefined human death as irreversible brain death, so that after irreversible brain failure it became permissible to stop further therapeutic efforts without being exposed to the reproach of negligence. That paved the way for an acceleration of the development of transplantation medicine. It is now well established and by the end of 1994 436,000 transplants had been performed worldwide. Two thirds of these were kidneys and the remainder heart, liver, pancreas and lungs. But the acceptance of transplantation decreased from 90% to 65% amongst relatives of brain dead patients. The case of the "Erlanger baby" may have contributed to this.

5 Donor organ failure has been on the increase since 1992 and the demand for organs is growing. To meet this demand work began on developing xenotransplantation (animal to man) initially from baboons and later from pigs. The risk of transmission of pathological agents, particularly viral contamination by the xenograft was recognised at an early stage.

6 A very important contribution to the development of xenotransplantation medicine arrived with the advent of genetic engineering. The aim of basic research in this area is to overcome the problems of the immune barrier. The following examples demonstrate this:

- UVB-irradiated rat hepatocytes intraperitoneally transplanted into rabbits with experimentally induced liver failure resulted in a survival ratio of 60%.
- Nicotinamide treated pancreas cell cultures of transgenic mouse insulinomas (MIN-6-cells) secrete insulin depending on the glucose concentration. The aim is to make a bio-artificial endocrine pancreas.
- With the aim of eliminating xenograft antibodies, a total blood exchange with PHP before xenotransplantation of a guinea pig heart to a rat was done and the survival time was significantly prolonged.
- In an in vitro model, pig kidneys were perfused with human blood. In the xenogenic circulation the following changes attributable to the rejection reaction (HAR) were noted within minutes: fresh capillary thrombi, fibrinoid glomerular capillary necrosis,

diffuse tubular parenchymal lesions and oedema with haemolysis. Concentrations of the following increased: LDH, free plasma haemoglobin and potassium. Also a progressive activation of the complement system with increased plasma concentrations of C3-products and terminal complement complex (C5b-9) occurred.

Since 1992 transgenic pigs have been bred to slow down the hyperacute immune-reactions. Two human genes from the complement activity regulator group (RCA) were inserted into the genome of pig ova, which, once re-implanted, led to transgenic pigs. Perfusing their hearts with human blood in vitro was not followed by hyperacute rejection.

The achievements in this field of research are immense and deserve much respect. But the concepts behind them are of a materialistic-reductionist character. They belong to materialistic determinism which has social implications. Several phenomena and facts make this philosophy uncertain. Living substance can be manipulated, but not created. Therefore the chemical explanation of life by DNA cannot be true especially as DNA depends for its structure and function on numerous specific enzymes. Also, new information can be integrated into its structure. The genome as an information carrier is not identical with the information itself, which belongs to a higher, immaterial level.

The achievements of modern science divide mankind into camps of pro and contra. At the same time one must go to increasing lengths to develop skills in science and it becomes more and more difficult to see it for what it is. There is therefore an ever growing need for us to develop our powers of judgement. This involves unfolding a way to knowledge that will help us orientate in this rapidly expanding field of gene technology and xenotransplantation.

Hitherto, reliable ethical principles have been eroded. Consensus conferences (e.g. brain death = death of the human being) are bad substitutes for such principles. How do we find criteria for judging and acting in these fields? How do we develop new ethics? This cannot be by statutes which only serve the public interest. No, ethics can only be developed on the basis of free individual knowledge (cognition) which does not have to be limited to Kantian philosophy (cf. Truth and Science, Rudolf Steiner). On this basis, open-mindedness and the will to perceive the scientific phenomena and techniques together with their underlying concepts is required. Only then can individual ethical judgment and decision became a basis for deed. One problem however is the rigidity of our concepts. But the achievements of gene technology should not be deprecated. Its roots are part of our own being. We ourselves are responsible for its products and have to take note of that responsibility.

To recognize the different levels of reality of modern science one must go deeply into the subject. This way reflects as it were the way into our own soul. The pain of the process of attaining knowledge of what has become visible-invisible can spur us to recognize in ourselves the "dragon" and the "angel".

Peter Braidley discussed different questions regarding xenotransplantation out of his experience in this field of work. Human to human transplants function satisfactorily, but the demand for organs is much higher than the supply. So there is a clear need for xenotransplantation.

The question as to whether a transplantation is acceptable in principle has to be answered with yes and no. It depends on many considerations including the individual case and the attitude of the patient. There is a large framework of conditions and requirements. It stands to reason that full information and freedom of decision is guaranteed, but social implications also play a part. There are rules regarding the medical appropriation of a patient. Moreover if the patient is too well, they will not receive a new organ. Special regulatory procedures exist and outer conditions such as time, funding and availability of manpower etc. play a more or less important role. The aims are in general to benefit more patients and to give them a higher quality of life. The patient must feel better afterwards. The means to achieving this whether by human-to-human transplants, xenotransplants or artificial devices makes no difference. Xenotransplantation for hearts and kidneys will be available in the next few years. But also social aspects play a decisive role. Questions arise about the new social dimensions created and about the relation of man to animal, about keeping the animals etc.

Other questions arose from the participants which were difficult to answer: What is dying – something timeless taking only a moment? Should we prolong life indefinitely simply because we can? What makes a man a man or a pig a pig?

Changes in the personality of the patient after transplantation have not been observed. But psychological feelings and dreams are not unusual. Therefore the patients get psychological care before and after transplantation.

The mortality rate immediately after kidney transplantation is relatively low at 1%. Life expectancy after transplantation is about 10 years for 50% and 5 years for 63% of the patients. They have to take immune-suppressives for the remainder of their lives.
The distinction between preventive medicine and transplantation was felt to have only a relative meaning because of the unique situation of each patient.

A question about how we can overcome the problem of the shortage of organ donors was discussed. Three approaches were identified: firstly through education, meaning early identification of donors and solving the problems of obtaining consent; secondly, through optimal medical management of the donor (already brain dead); and thirdly through artificial organs.

The question of consenting to a transplantation is in the first instance a purely individual decision. But it has social implications and this makes it very complex. On the other hand transplantation medicine has become an established method. Can we stop it? Do we know its influence upon society? Each patient has a right to medical care including transplantation. Here also the problem of criteria arises: e.g. age, smoking, alcoholism, obesity, diabetes, cancer etc. The patient's social background has no influence. Money comes into the decision-making in USA but not in the UK. The whole complex of questions and considerations, many not purely rational in nature, have to be taken account of.

The choice of animal organs for transplantation their associated risks for humans from the physical point of view were discussed. Size and function must be adequate, which is the case with baboon and pig. A pig heart last 3-11 years. Very much research has been done on the pathogenesis of hyperacute rejection, on the interaction of xenoreactive antibodies and target antigens on cell surfaces. The research has produced initial results in the form of transgenic pigs with one human cell surface protein. 2,500 ova were required to produce 300 piglets of which 49 had the human protein on their cell surfaces. Subsequent generations retained the same feature following standardized crossing techniques.

Account has to be taken of the danger of infection, but this can be tested for. Also there is a great deal of research in this field.

Peter Braidley pointed out that the transplantation technique is effective and safe and therefore provides a potential benefit for patients. Against this the point was raised that nature would be exploited by xenotransplantation. Furthermore, although the physical point of view mentioned above is fundamental, it is not by itself sufficient or adequate because the human being is more than a physical body. Someone asked whether we know the interrelationships between the level of the physical body and the "higher" levels which demand a different scientific approach. This question remained unanswered.

Summarizing the different aspects: it can be concluded that there were not enough viewpoints to get a satisfactory overall picture. Whilst patients are human beings, the research was done on animals. In this context then, what relevance does human biography or even the process of dying have? Little is known of what would be taking place in the body and soul of a man with a pig's heart. In the fields of molecular biology and genetic engineering such questions still have no meaning nor are they raised, yet they are relevant for the patient in relation to his sense of self and his individual decision about transplantation.

Answers to these questions will have a retroactive effect on transplantation medicine. More research is also necessary in these fields, especially at the psychological and spiritual levels, in order to develop viewpoints beyond the purely physical. It would contribute to setting the goals for genetic engineering and xenotransplantation.

In this context a suggestion was made for doing some research in the form of an exercise. One should try to imagine in as detailed a way as possible how a living being arises. A picture will develop, which should be concentrated upon for a period of time. Then it should be put aside and instead one should try the same exercise on a transgenic animal, holding back any judgment. Another picture will arise. This exercise should be repeated for days or weeks.

We are free and have the wherewithal to carry out transplantations, but moral development has not progressed in proportion to our freedom. The moral level is connected with the social implications of our deeds. In addition there are questions of safety.

Further problems have to be considered: What happens to evolution when the species barrier is crossed? Transformations across this barrier are new and the consequences not yet known. We are gaining experience in this field only by trial and error, so the valid question arises as to whether we should base our scientific work only on such experience. If we do not see through the whole complex of implications the answer should surely be 'No'.

The criteria for our way of acting are always derived from our own knowledge and we have to be honest when face to face with the patients.

Reference

Pertinent literature is available on request by J. Jungermann.

25 Embryo, heredity and DNA

WORKSHOP

JAAP VAN DER WAL
Embryologist
Department of Physiotherapy, Hogeschool Utrecht
Bolognalaan 101, NL-3584 CJ Utrecht

PETER GRÜNEWALD
General physician for mentally disabled children
14 The Deil, Westbury-on-Trym, BS9 3VD Bristol, UK

TROND SKAFTNESMO
Biologist
Steinerskolen
PO Box 467, N-5501 Haugesund

Abstract
The study of embryology has developed from a phenomenological, descriptive and comparative approach towards an experimental approach, which is more and more carried out on a molecular level. On this way towards the 'infinitively small', there has also been a shift in the level of explanation: The current model of morphogenesis is exclusively dealing with concepts emerging from a molecular level, such as 'position information' and 'gradients of morphogens'. In the workshop-discussions there has been expressed a doubt whether this approach can solve the riddles of morphogenesis. Especially there has been pointed out that the chemical agents named 'morphogens', such as for instance *retinoic acid* or the protein *bicoid*, are by no means the creators of form (morpho-gen = form-origin). Instead they must be acknowledged as door-openers for certain form-processes: they are necessary but not sufficient conditions (i.e. not 'causes') for the morphogenesis.

The role of DNA in development

JAAP VAN DER WAL

For many people nowadays the outcome of the 'debate' about *nature or nurture?*, as far as the structure and function of an organism (phenotype) is at stake, is more and more in favour of the former. Genes ('nature') are considered to be the morphochemical substrate of a program that governs the organism to its phenotype. Environmental conditions ('nurture') only play a moderating role in the genetic program. In this view the results of genetic manipulative procedures and experiments are considered to 'prove' the 'one-way-dogma' of molecular biology which represents the view that the genetic program is the cause of the phenotype. This radical view overlooks the fact that genes or genetic program may be a necessary but never are a sufficient condition for the appearance of an organism. In particular in developmental biology it becomes evident that the phenotype of an organism at least results form *preformational* and *epigenetic* conditions and that, like Waddington stated, properties are always the outcome of a 'joint action of environment and genotype' (Waddington, 1975).

In the embryonic development this is illustrated by the process of *differentiation*. For example in the development of most mammals the first differentiation from the stage of *morula* – in which the cells are phenotypically and genetically more or less similar – is represented by the differentiation into an *inner cell mass* (in the human: *embryoblast*) and an *outer cell mass* (in the human: *trophoblast*). There is proven evidence that this differentiation is initiated by the different relationship, which the outer and inner cells of the morula have to their nutritional environment, resulting in metabolic changes within the cell and this again producing intracellular factors that moderate the state of activity of the genes within the nucleus of the cells (Hinrichsen, 1990). Differentiation is a process from outside to inside. In terms of modern molecular biology it means that genes are 'activated', 'silenced' or 'repressed' by 'signals' mediated by 'messenger proteins' originating from the environment of the cell at stake. It also has become clear that the signal neither the response on the genomic level needs to be specific. Like in the *induction* experiments of the first decades of developmental biology (Hamilton, Boyd and Mosman, 1971; Wolpert, 1991; Wadddington, 1957; Weiss, 1986) the signal may be a specific enzyme or protein but its response might be mimicked by many other substances. In the process of differentiation the statement of the human embryologist E. Blechschmidt still is true: 'Genes are not active principles, they re-act' (Blechschmidt, 1976).

The processes of differentiation indicate that the definition of a gene as a sequence of nucleotides on a chromosome should be replaced by a more functional one. The gene rather is to be considered as a task that (at least) the cell has to perform. The substrate of the DNA is a necessary but not sufficient condition for differentiation and never can be

pointed out as the 'cause' of the process. Differentiation has no cause, it is cause itself. DNA is not an active performing principle in a cell or an organism, it is reactive and only to be understood in the context of a (living) environment. The human genome can only be expressed in a biological environment of human quality. The myth of Jurassic park has proven to be untrue: we will never be able to restore dinosauric life as long as a 'dinosauric environment' (at least a 'Jurassic' egg or ovary) fails us.
Differentiation is ruled by many factors that are not genomic at all, which does not mean they are not hereditary. Some of those factors are:

- *Age* of the embryo. In young embryos complete organisms might develop from all constituting cells. Later on this potency is restricted.
- *Location* in the embryo. During a certain period of time – the so called 'induction window' – a cell that already has been differentiated, can be replaced experimentally to another area or place in the embryo after which event the cell will differentiate according to the pattern of the related area in the embryo, i.e. its new environment.
- The grade of differentiation in the developing embryo grows *in time*. When a certain stage of differentiation has been reached, development can only be continued in *that* given direction: 'there is no way back'.

This all confirms that differentiation is a process in time and form outside to inside and that considering 'hereditary' as synonymous with 'genetic' or 'genomic' represents odd reductionism. Modern biologists stating the existence of so called *morphogenes* as the substrate of differentiation of spatial relations in the developing embryo, are only right if they consider those genes as being 'tasks units' of the organism. However many of those scientists persist in a pure morphological or structural definition of genes as circumscribed areas of DNA and claim they have found the DNA for 'morphology'. It is this ideology that claims *hox genes* to be evolutionary stable and seduces to state that morphogenes in diverse organism might be the same or similar and are presented as such to the lay public (Davis, 1995). In this way the same error will rise again that was overcome in developmental biology when the Biogenetic Law of Ernst Haeckel was proven to be wrong in the sense that two developments might be homologous or analogous, but that this does not mean that the underlying processes in the individual ontogenies are the same! *Knock out* experiments with hox genes from Drosophila introduced in mice will prove that, if there exists something like a morphogene for the eye, it will at least 'provoke' in the mice organism a mice eye and never a Drosophila eye. Instead of claiming to have found the 'origin' of body plans in the form of hox genes and morphogenes, a more modest definition of the revealed processes like Nobel price laureate Mrs. Nüsslein-Volhard handles, as she states the existence of 'gradients that organize embryo development' (Nüsslein-Volhard, 1996), would avoid such fatal, i.e. principal mistaking. The entity of life and development at least is the organism, not the gene(s), in particular not the genes as defined by the regular molecular biologist i.e. as a certain localized sequence of nucleotides in a DNA string! In the process of differentiation it becomes clear that the genome at least plays a secondary role in the events of life. Awareness of the fact that the gene is a 'fuzzy concept' (Fischer, 1997) would never tempt to statements as 'DNA is the molecule of life' (Watson, 1968), giving base to the philosophical misthinking of the *genetic essentialism* (Lindee, Nelkin, 1997).

Usually morphogenesis is considered as a biological process leading to or continued in physiological function and/or psychological behaviour in the later phases of the development of the organism. Form is considered to be the primate, function and behaviour as secondary. A phenomenological approach in biology however does not discriminate between form and function or between form and behaviour. 'The embryo functions in forms' (Blechschmidt, 1976), or 'a living system exhibits behaviour as growing organism' (Weiss, 1986), are statements which direct the attention to the biological behaviour of the developing organism itself as it is growing and developing. Form as a function of the system. This requires an organismic approach which at any moment of the development takes into account the *Gestalt* of the organism.

This is demonstrated by means of some examples taken from the human embryological development. The transition from the first week stage of 'blastula' to the second week phase of 'embryocyst' represents a totally different relation between the embryo and its environment. In the first week the embryo is more or less independent and is to be considered as a rather autonomic 'introvert' system, ruled by the process of reduction cleavages. Would this process be continued the embryo would end up as a mass of cells with the high *nucleus cytoplasm ratio* known from spermatozoic cells. The event of nidation (implantation) represents a complete reversal of the relation of the embryo to its environment. From then on the embryo is exhibiting, physiologically as well as morphologically, an intense tendency towards periphery. Radial outgrowth and threshold passing biological behaviour (formation of an active *syncytio- and cytotrophoblast*) with related peripheral influencing of the maternal organism (*gonadotrophin* excretion) represents an 'extrovert' relationship with environment. The rather 'closed' organism of the first week changes dramatically into an 'open' organism. This tendency on its turn is self-limiting. If the embryo persists in this behaviour it will end up as a so called 'wind egg' in which trophoblast (or *chorionic vesicle*) – and therefore pregnancy! – is present but an ('inner') embryo is absent. Again an essential turning point in biological behaviour (*Gestalt*) is exhibited when in the middle of the third week of human development the direction of the processes is turned into the reverse. With the appearance of a capillary blood system resulting in the formation of a primordial heart in the centre of the embryo (in that sequence!), the dynamics of the development are directed from periphery to the centre. This starts the individualisation of the 'inner' embryo (the now threefold layered germ disc) from the 'outer' embryo (the later chorionic vesicle). This process of emancipation for the moment ends with the *delamination* at the beginning of the fourth week of development and represents the principal dynamics of the process of birth. With that event the emancipation or individuation, started morphologically in the third week of development as 'growth behaviour' is then so to say performed at the physiological level.

The aim of this short example of *dynamic morphology* is to show that description of the biosystem as a whole and taking form or morphology into account as behaviour of the organism reveals principal or essentials properties of the organism, that would remain unnoticed by the one who only takes into account the events on the level of the 'DNA program'. A dynamic morphology includes the regulatory events on the genomic level as necessary but not sufficient conditions of life and development. The reverse is not true.

The following report is a summary of the contributions of the workshop-leaders, Trond Skaftnesmo and Peter Grünewald, as well as the main viewpoints which were expressed during the workshop-discussion. Presentations and discussion are incorporated in the following summaries.

Organising the embryo

TROND SKAFTNESMO

In 1995, the Nobel Prize in medicine was given to three laureates: *Edward Lewis, Christiane Nüsslein-Volhard and Eric Wieschaus* – for their discoveries concerning 'the genetic control of early embryonic development', as the Nobel Assembly put it. We may come back to the reality and impacts of their discoveries later on, but firstly we must have a brief look at their research project and how their findings are interpreted by themselves and their fellow molecular geneticists.

Their 'experimental system' (to quote the Nobel Assembly) has been a living organism, the well-known fruit fly, *Drosophila melanogaster.* In this insect they were able to identify and classify a small number of genes that are of key importance in determining the body plan and the formation of body segments. They treated flies with mutagenic substances so as to damage (mutate) *Drosophila* genes at random. In this way they could decide which genes, when damaged, would cause disturbances in the formation of the body axes or in the segmentation pattern.

This research has further revealed some of the metabolic pathways by which the genes seem to control the early embryonic development. The main principle in action seems to be polarity established by means of chemical gradients. As an example we can mention the gene named *bicoid*, which is involved in patterning the anterior end of the embryo. This gene, which is one of the first to be activated during the development of the egg, encodes a protein – simply called the bicoid protein. When the drosophila-egg is laid, and before cell formation, this protein begins to be synthesised at (what is to become!) the front end and then diffuses along the egg. As the protein is continuously produced at the front end, and as it is gradually metabolised along the path to the rear end, it will set up a stable concentration gradient along the postero-anterior axis with the top at the front end. This gradient, so goes the story, is now able to 'tell' the cells their position along the main axis and thus, where they for instance should 'form' the border between head and thorax. It is stated, that by setting up this gradient the bicoid gene is controlling the development of the anterior part of the body. At least it is shown to be enough to mutate the bicoid gene to develop a larva lacking both head and thorax. As a control experiment it can be shown that such mutated eggs can be rescued by injecting cytoplasm taken from the anterior part of other, normal eggs: when the bicoid gradient is restored, the head/thorax-development is again fully normal.

Figure 1. The Drosophila-egg and its bicoid gradient.

The gradients set up by the bicoid protein and two other proteins with similar function, is supposed to determine the initial segmentation-pattern of the larva. The process of making segments however is considered to be only the first half of the job. The second half is to make them different, to give them a separate identity. This is taken care of by genes standing in a lower hierarchy, *the homeotic genes.* (If it is stated here 'taken care of by the genes', it only means that when one destroys (mutates) a gene of this kind, one will confuse the formation, the differentiation of certain segments in a certain way.) The most surprising discovery concerning these homeotic genes was perhaps that their order along the chromosomes was the same as their expression along the body axes: the genes expressed at the anterior body segments (the head) was sited first in the gene complex, the genes expressed at the posterior end was placed on the opposite side of the complex and so on. There has also been found a similar genetical controlling system for the dorsoventral (top-bottom) polarity, though the gradient is set up in a slightly different way. And finally, there are found similar homeotic genes in vertebrates. In mammalians

there have been found four such gene complexes (clusters) known as the HOX genes.
The question arising is; Are these experiments and findings the ultimate proof that 'the gene-program approach' to morphogenesis is 'true' after all? Most researchers of the field seem to have taken this for granted. So says for instance the mentioned Nobel Prize laureate, Christiane *Nüsslein-Volhard*: 'Concentration dependence and combinatorial regulation together open up a versatile repertoire of pattern-forming mechanisms that can realize the designs encoded in genes.' The belief that form or design is encoded in the genes is – as we can see – manifest. But Nüsslein-Volhard is a cautious scientist, and it's interesting to see how she actually expresses her view in a modest way: 'Concentration gradients ... open up', she states, 'for the "versatile repertoire" of design-coding genes.' But this should be read as: The concentration gradients is just the inducing side, the *signal-side* of the morphogenesis, and the *response-side*, the making of the form, – well, that is the 'versatile repertoire' which is yet to be uncovered.
The difference between signal and response is very clearly expressed in the example of the chicken limb bud. To make a long research story short, for several years there has been a search for the position-signal that 'tells' the cells in the limb bud of a vertebrate where it is placed and hence what it is supposed to become. Actually, there is a search for several signals, since we have at least three axes to consider. The proximo-distal axis, the anterior-posterior axis and the dorsal-ventral axis. Whilst the other signals are still to be

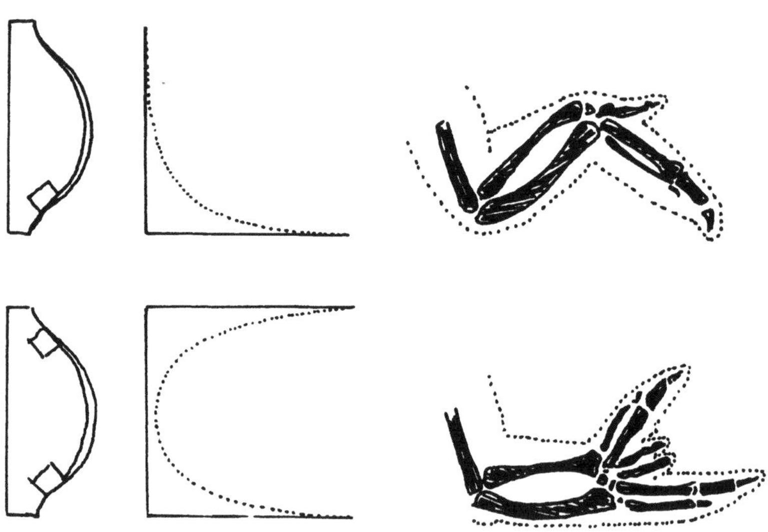

Figure 2. The chicken limb bud and its gradient of retinoic acid.

discovered, the signal across the limb bud from the posterior to the anterior side was finally revealed by Lewis Wolpert and his research team. It was found in the chicken limb bud and the gradient is set up by *retinoic acid* which is a derivative of vitamin A. At the posterior margin of the limb bud there is a small region called *the polarising region.* In this region there is a maximum of retinoic acid, whilst there is a minimum at the opposite end of the limb bud-tip (The thumb end, so to speak, even though the chick lack the thumb – it has only the three digits in the middle). The following experiment demonstrates the impact of this retinoic acid gradient: A hole is made in the eggshell and a small cube of cells from the posterior margin, containing the polarising region, is cut out and grafted into the anterior region of another chick-embryo. This chick-embryo will now have one polarising region on each side of the limb bud and thus have two maxima of its retinoic acid-gradient. The gradient will be U-shaped and consequently, the limb bud will develop into a monstrous limb with a double set of digits; a mirror-image of the 'chicken-hand' will form on the anterior side.

The question is raised whether it is correct then to name this chemical substance retinoic acid a *morphogene* (that is 'form-maker' or 'form-origin'), which is done by Lewis Wolpert and other researchers of the field? It is quite clear that it is not the retinoic acid that 'creates' or shapes the form, the digits – it's job is simply to be the guardian at the door, that *opens up for* this morphogenesis. And this can be conclusively demonstrated by the fact that even in the limb bud of the mouse embryo, there has been found a similar gradient of retinoic acid. But the result is definitely a mouse limb, not a chicken limb. If you take a piece of tissue from the posterior margin of the mouse limb bud and graft it into the anterior margin of the chick limb bud, you'll get a chick-monster hand, not a mouse monster-hand and vice versa. Finally, there is a big difference between a birds wing and its foot, but the 'morphogene' is the same i.e. retinoic acid. So it is clear that it is the responses on this so-called morphogene that make them different and, likewise, make the digits of birds, mice and men different. But the response-side is vastly more complicated than the signal-side. A network of interactions between genes and proteins appearing as activators or repressors exists. The proteins involved are just as much (or less) 'causes' as the genes and vice versa.
If we go back to the statement that the response is nothing but the making of the form – which was actually the problem to be solved – have we after all come any closer to a solution for it? Or have we just moved further away from a solution of the problem, since we perhaps are using all our effort to search for the solution in a domain where it cannot be found? What, after all, is or means form in the organic world? Do the presumed genetic program have a 'concept' (representation) of the form or, if it has not, – which is probably the correct answer to that specific question – how can it then 'make' structure, 'create' form?
Lewis Wolpert has a favourite metaphor (analogy) to this question, when he states that

the gene program does not contain any blueprint or model of the body, but it contains instructions on how to make it. The analogy is *origami*, the art of making different spatial forms (animals and so on) by repeated folding of a sheet of paper: There is no need to have any image of the end product i.e. the form as such, if you have an instruction which tells you how to fold the sheet, step by step. By analogy, Wolpert states, there is nowhere in the genes any representation of the body or any organ, only mere instructions for stepwise folding of the sheet. Following however the logic of the same analogy, we could ask in return: Would there be any art of *origami* if there where no instance where the form was represented? It is correct that you can make a form by mere stepwise instructions; but can you also make the instructions blindly?

'The central dogma' of molecular biology has been falsified by modern genetics

PETER GRÜNEWALD

How should we understand the molecular-biological relationship between nucleus and cytoplasm, between DNA and enzymes (proteins)? I will briefly try to examine the idea that the synthesis of proteins can be viewed upon as a mirroring of the DNA-structure, mediated by RNA, and that the protein can be considered as the physical instrument for the life-processes (metabolism, growth and reproduction).

1. We have to regard DNA and protein as *carriers of information* and not as *the information itself.* Proteins (enzymes) are catalysts of biochemical reactions in the cell. They control and carry out these reactions, but they do not 'take the initiative' to bring them about. DNA and proteins are both necessary tools for the living, cellular processes, and they are also constraints for these processes but they are not their creators i.e. they are necessary but not sufficient conditions. The actual properties of these tools have a decisive impact on the variability of the life-processes. The formation of the cell as such and the totality of the cell-processes cannot be deduced from its DNA/protein-equipment (Grünewald, Wilmar, 1992). From what source then does this formation emerge?
2. The flow of information does not go exclusively from the centre (nucleus/DNA) to the periphery (cytoplasm/protein), but the metabolism of the periphery has counter-impact on the DNA-activity – in particular through the 'transcription-factors', which are certain enzymes which activate or repress the expression of the genes. (Heusser, 1990).
3. There is no possibility whatsoever to predict the *Gestalt* of an organism or an organ from its genetic equipment (See *introduction*).
4. Thus, the gene-model of recent genetics has no positive explanation value concerning the riddles of morphogenesis and the source of organic forms.

Instead of taking the contemporary gene-model as the sole bases of considerations on morphogenesis, it seems fruitful to complement it with the model presented by Prof. C.H. Waddington of Edinburgh, who has developed the idea of *the epigenetic field* (Waddington, 1957). This model explains morphogenesis and metamorphosis of organs as a result of the interaction between DNA and a morphogenetic field. This morphogenetic field penetrates and surrounds the developing organism. It seems to organise cells and allows the regulation of the activity of specific genes, which are particularly important to the specific activity and formation of this specialised cells. Specific parts of the DNA are thereby activated or suppressed and cause the specialisation of organelles regarding form and function.

This model seems to be supported by embryological experiments, of which some have been well known for decades (Weiss, 1986). As an example we will mention the experiments carried out by O. E. Scott in 1935. Scott transplanted specialised cells of a salamander-embryo from one organ onto another and showed that in early phases of development, the transplanted cells changed morphology and function and adapted completely to the new environment. The specific genetic program of the specialised cell was changed through interaction with a different environment. This indicates, that organisation, form and function of cells are not just determined by their (omnipotent) genetic make-up, but through interaction between DNA and the organic environment of the morphogenetic field with its organising, form-creating and function-determining forces.

Conclusive remarks

The existence of a 'time-space-body' – the indication 'body' in fact is to anatomical or structural for what is meant here and might be replaced by 'field' – as the real quintessence of the organism has not been ruled out by modern genetics and molecular biology. Even more, the hypothesis of such an immaterial quality includes the facts of modern molecular biology but directs the attention to a hierarchically higher level of living beings.

References

Blechschmidt, E. (1976) *Wie beginnt das menschliche Leben?*, Christiana-Verlag, Stein am Rhein, Switzerland.

Davis, S. (1995) Invasion of the Shapechangers, *The Ronald Grant Archive*, 28 October, pp. 30-135.

Fischer, E.P. (1997) The archetypal Gene: The open History of a successful Concept, in: this volume.

Hamilton, Boyd and Mossman (1971) *Human Embryology*, Heffer and Sons, Cambridge.

Heusser, P. (1990) Das zentrale Dogma nach Watson und Crick und seine Widerlegung durch die moderne Genetik, *Der Merkurstab, Heft* 3.

Hinrichsen, H.V. (1990) *Humanembryologie*, Springer Verlag, Berlin, New York, London, Paris, pp. 87-91.

Lindee, M.S., Nelkin, D. (1997): *The cultural Powers of the Gene: Identity, Destiny and social Meaning of Heredity*, in: this volume.

Grünewald, P. and Wilmar, F. (1992) *Vererbung und Genetik*, Verlag am Goetheanum, Dornach, Switzerland.

Nüsslein-Volhard, C. (1996) Gradients that organize Embryo Development, *Scientific American* **39**, August, pp. 38-43.

Waddington, C.H. (1957) *The Strategy of the genes*, Allen and Unwin, London.

Waddington, C.H. (1975) *Evolution of an Evolutionist*, Edinburgh University Press, Edinburgh.

Watson, J.D. (1968) *The double Helix*, Penguin Books, London, New York.

Weiss, Th.J. (1986) *Embryogenesis*, Floris Books, London.

Wolpert, L. (1991) *The Triumph of the Embryo*, Oxford University Press, New York.

26 Meeting the person behind the scientist involved in gene technology – summary of the round table discussions

ROUND TABLE PARTICIPANTS:

HERMAN DE BOER
Biochemist
Institute of Chemistry, Department of Medical Biotechnology at Leiden University
PO Box 9502, NL-2300 RA Leiden

BRIAN GOODWIN
Biologist
Department of Biology, Open University,
Walton Hall, Milton Keynes, Bucks MK7 6AA, UK

BARBARA HOHN
Plant molecular geneticist
Friedrich Miescher Institute
PO Box 2543, CH-4002 Basel

ALBRECHT LINDEMANN
Medical doctor
Dept. Haematology/Oncology, Freiburg University Hospital
Hugstetterstraße 55, D-79104 Freiburg im Breisgau

PIA MALNOË
Microbiologist
Institut fédéral de recherche agricole de Changins
CH-1260 Nyon

REINHOLD SALGO
Physicist and patent attorney
Töbelistrasse 88
CH-8635 Dürnten

CHAIRMAN: BAS PEDROLI
Ecologist
Delft Hydraulics
PO Box 177, NL-2600 MH Delft

Reported by Andree Dargatz (d – Hamborn),
Edith Lammerts van Bueren (nl – Driebergen) and David Heaf (uk – Gwynedd)

Introduction

The round table discussion took place over two evenings and involved five scientists who are experienced in fields related to heredity and gene technology. They commented on their scientific and personal motives for the work they do. On the first evening the chairman focused his questions on the success side of the story and on the enthusiasm which motivates these individuals in their work in the field of genetic engineering. On the second evening, the dilemmas they face in their work with genetic engineering, the ethical questions and their future goals were discussed. The contrasting perspectives of the participants in these discussions gave a many-sided view of the scientific, ethical, social and personal approaches of people in biotechnology and its associated disciplines. All panellists described vividly the events in their biographies when they became enthusiastic about the efforts of genetic engineering.

The scientist's very first moment of enthusiasm for the subject

The discussion started with a request to the panel members to describe the first moment in their lives when they became enthusiastic about genetic engineering. Each responded with different and very biographical accounts making it clear that the majority of such moments arose in combination with fundamental research on DNA and genetic engineering. For example, in the 50's when he began his Ph.D. studies, Goodwin was excited by Watson and Crick's extraordinary discoveries of the DNA helix and especially Monod's bacteria research. The kindling of his enthusiasm for biology could not be defined by a single moment but was by his reading in scientific journals and by the climate of excitement engendered by scientists visiting the University of Edinburgh. At that time his subject was going through a transformation into molecular biology and molecular genetics. With his mathematical background he was especially fascinated by the dynamic side of genetic control mechanisms.

Hohn spoke of her enthusiasm which arose while finding out how bacteria insert gene sequences in plant cells. Together with her husband at the Basel Biocentre she was trying to understand the process of virus maturation. One day she began an experiment which involved mixing DNA, proteins and other substances in a test tube and the following day she found something which had been growing and replicating. In vitro packaging of viral DNA was born. For her this was a very exciting experience and lead to a useful procedure for studying virus maturation and how to make DNA infectious, a process which is used to this day in genetic engineering.

For Salgo, as physicist and patent attorney, it was a case of his enthusiasm for the questions gene technology was asking. This began a few years ago when he became involved with the question of the oncomouse patent in Germany.

The very first moment that De Boer realised the direction his career was to take is still very clear in his memory. At the time he was a student in the USA. One day he was standing under a tree with a copy of the journal Nature in his hand, reading about a company which had managed to produce bacteria containing a synthetic gene for human insulin. He was excited by the idea that the human body produces many potential medicines which can be artificially produced with the same technique for use in healthcare. Within a week he had applied for a postdoctoral position with the company and worked in its stimulating climate for eight years on human growth hormone expressed in bacteria. After that he returned to the Netherlands to work with gene technology in cows.

For Lindemann, a physician, it was not a matter of just one moment but more a slow growing into the field with its new discoveries that genes regulate types of cancer such as leukaemia; with its possibilities for more precise diagnostics; and with its potential for genetic treatment of the disease.

Experiences prior to taking up science

On the question of whether any experiences before taking up scientific studies, for example in childhood or at school, had influenced enthusiasm for scientific involvement, Goodwin recounted how two elements had been of importance. He grew up in Canada where he was able to develop a very close relationship with nature and for a time thought he would become a forest ranger. But at the age of 15 he read a biography of Madame Curie and experienced the romance of discovery in science. That was the life for him!

De Boer's youth as a farmer's son in a little Dutch village did not at first influence his later laboratory work in the USA. But after coming back to the research in the Netherlands in 1987, the experiences of his youth came back to him. He wanted to combine his interest in producing biomedical proteins by transgenesis with his childhood passion for dairy cows. And that is how he became the 'spiritual father' of 'Herman', the world's first transgenic bull. Working in this commercial context he could still feel to some extent a farmer, as he was one of the few in his team who knew how to deal with raising cattle.

Hohn now sees that her stubbornness in continuing her chemistry studies against her mother's wishes helped greatly in her later career to maintain faith in her experiments even when they did not yield immediate success. Scientists must be stubborn in order constantly to improve both their models and their experiments.

Salgo's first interest in science as a child was in discovering what lay behind the dashboard of the family car or exploring the insides of household machines, but his serious interest in the questions of science arose when he learned about the duality of wave and particle views of, for instance, the electron.

Randomness versus purposefulness in scientific research

The chairman asked: is modern science something which develops randomly or purposefully? Goodwin responded that only about four years ago he felt impelled to become engaged in the discussion about gene technology in order to balance out a deliberate tendency he had observed in biology. This was the tendency of his subject to move away from the ecological and macroscopic/holistic approaches to understanding the organism as a whole in favour of an overly rapid development of the microscopic or genetic reductionist view which involves understanding an organism in terms of genes. His concern had always been to grasp the context in which genes operate. Although he is in favour of biotechnology, he feels that its development is proceeding too fast to be properly regulated in the way, for instance, the pharmaceutical industries are. Thus, he argues, we have to develop an international biosafety protocol and this involves trying to overcome the positivist split between science and ethics.

According to Hohn, science develops mostly at random. She illustrated this by reference to basic research on gene transfer by bacteria that cause cancer in plants which then led to this principle being exploited to put any gene of choice into another organism. She does not regard the random approach as necessarily ideal but it has potential if scientists are vigilant about what is going on around them. In basic research new connections can suddenly come to light.

To the question of whether science develops at random or purposefully, De Boer replied that he thinks neither of the two is solely the case. There are many examples where random research projects from different teams can suddenly turn into research with a very definite purpose when the results are combined. He illustrated this with reference to work on plasmids in the USA becoming combined with Swiss work on enzymes that cut DNA and eventually leading to making bacteria that produce medicines. But these days academic research is very intertwined with companies who want to make it more purposeful, so much so that random research inspired by the scientist is suffering badly. It can only proceed if there is ample money for it. And where can you find that these days?

Safety issues

Lindemann asked Goodwin to enlarge on his concern about the safety of gene technology by focusing on the issues which are specific for gene technology itself. To clarify his question Lindemann added that he knows many safety problems, for instance in the field of vaccines, that are not solely attributable to the fact that gene technology is involved. Goodwin answered that biotechnology applied to food is not regulated as stringently as are pharmaceutical products. For instance the food products are not appropriately labelled. So if something goes wrong with the product there is no possibility of tracing the origin of the damage. Whenever you change the genetic composition of an organism the probability is that you will get metabolites you did not expect. Concerns like this justify asking for safety protocols equivalent to those for pharmaceuticals. But there is an important difference to be considered to do with containment. With transgenic plants you need to be certain that the organism is stable, that no horizontal gene transfer can take place and that there won't be ecological damage. That is just basic biological common sense.

Here, a member of the audience asked De Boer, "Would you drink the milk from a genetically manipulated cow with the same relish as that from other cows?". De Boer answered in the affirmative but added that although it is true that if you make genetically modified organisms you may generate organisms that produce new peptides or that might be toxic, the same can happen with conventional breeding. But you always have to set up a selection program to select the best variety and you proceed with that one. De Boer is not convinced that there are more problems with recombinant DNA technology than we are already aware of in traditional breeding which also entails DNA recombination. Goodwin responded that the risk of destabilisation of transfer over greater taxonomic distances is a more serious problem than in classical breeding between related organisms. A consensus between all participants of the discussion was reached that destabilization by the unknown product is not only a problem of genetic engineering but for breeding and ecology in general. Goodwin pointed out that it was nevertheless urgent. According to Hohn, we should set controls whatever the method, classical or transgenic.

Another question from the audience focused on the commercial influence on research, for example the production and early release of a potato with more protein. We should not only concentrate on safety issues, but also ask questions about needs and commercial goals. In response to this, Hohn agreed that there must be a balance between the benefits of the research and the risks. Another commercial interest is the patenting of some new genetic research product, e.g. the oncomouse. Salgo drew a comparison between genetic engineering and the nuclear power research process in the 50's and 60's. He recounted how at a certain point the scientists involved considered the physics to be solved so the emphasis, strongly influenced by the flow of funds, was switched to nuclear chemistry. The mushrooming of biotechnology companies that began twenty-five years ago was, he said, also money driven.

Evolution

A questioner in the audience asked "If DNA is not the motor of evolution but rather a static 'registration office' recording evolutionary change, what is actually driving evolution? How does information come into it and how does it change?" De Boer responded that few would say that evolution is driven. The questioner added that if DNA tended to fixation, evolution would stop. For Goodwin this appears to be a paradox of stability and fluidity. We can resolve this paradox by attending to the different timescales on which the processes take place and to the dynamic attractors involved. To Goodwin, evolution is driven by three components taking place on different timescales: the fluidity of the genome (mutation/recombination); the effect of the environment; and the intrinsic symmetry-breaking properties of the whole organism as an entity, as a process giving rise to a complexity of form.

Lindemann commented that, for example, the stability of DNA might be disturbed not just artificially, but in nature by a retrovirus which triggers a tumour.

Discomforts and constraints

The discussion then shifted to feelings and scruples: "Don't the round table participants feel any discomfort about the technology?" Salgo admitted that he has feelings of discomfort, but does not know why. Taking our feelings as the only basis for judgement of solutions to such problems is a bit narrow. That, he felt, is why we were at the conference – to broaden our basis for judgements. De Boer admitted that he once felt uncomfortable in his work. He had generated a transgenic dairy animal for producing biopharmaceuticals. This was based on his personal idealism combined with some commercialism. A heated public debate of this issue arose and this made him rather uneasy because he found himself wondering 'Who am I to decide by myself to do this work?' Because of the public commotion a committee was formed. Politicians became involved and the end of the story was that it was ruled that one cannot do this sort of work unless there is a very good reason. To produce a biomedical protein was then considered to be a good reason. And that took away his uneasiness. What he had done was no longer resting on his own shoulders as inventor and developer of the technology, but became the responsibility of the whole society. That's why he always participates in any kind of discussion about this technology with whoever is interested. He thinks it necessary and wants to share what he is doing with others and hear their opinions.

According to Lindemann, the high rate of discomfort with genetic engineering occurs because we are related to living beings more than to other parts of nature, which would lead us to conclude that any change of DNA will have an effect on ourselves through our being related via DNA. This is a specific problem for genetic engineering. Lindemann defends his own involvement with this technology primarily because he is a physician and is interested in discussing possible uses of gene technology, for instance for recombinant insulin. To some degree it may be helpful and to everybody's benefit, but there are other issues that may be harmful. He feels we should discuss what should be done and what should not be done.

A questioner from the floor asked if any of the round-table scientists could imagine an application of the technology in which they would support a non-governmental organisation (NGO) campaigning against that application. Hohn can imagine very well such a situation because she felt we should not simply do whatever it is possible to do. It is a wonderful procedure but we have to use it correctly. We have very carefully to evaluate the benefits and risks. If substances arise in a transgenic plant which are too toxic for animals and not enough experiments have been done, she would oppose its use. One should judge case by case. There have been experiments that have been discontinued by companies because of concerns, e.g. allergenicity. Salgo has already directly supported an NGO specifically in its opposition to patents on living organisms.

The final question from the chair on the first evening was, "Do you always feel constraints or are you really free in your scientific work?" For Lindemann, who is not a scientist engaged in basic research, constrains are a necessity. But he added that in research which is not primarily applied, the scientist becomes free from constrains like money. De Boer stated earlier in the discussion that he experiences the public debate in the Netherlands not as a constraint but as a necessity. Goodwin stated that science cannot ignore the constraints of morality and the effects of genetic engineering on ecology. Feelings which tell us that something really important is happening deserve attention.

Second evening discussion: dilemmas

On the second evening Malnoë took the place of Lindemann who had to attend another congress. The chairman began by focusing on the dilemmas scientists encounter in their work life. He introduced this theme by giving an example out of his own work experience. He is responsible for an environmental impact assessment for a hydroelectric power plant in the Netherlands where it is looked upon as green electricity. One of the problems is preventing the fish from swimming into the turbines, but he was hindered from implementing his own mitigation measures because it was too difficult to justify these under the constraints of a restricted budget and demands for compromise. He experienced the dilemma of being in a consultancy firm as a scientist longing for more scope for fundamental research and really being able to work with particular ecological questions.

For Goodwin, the greatest dilemma is the separation of art and science, and the fact that scientific knowledge has become separated into many narrow fields. Working within such limitations in present day educational institutions is a real tension for him. He experiences himself as deeply imbued with this culture of separation and has difficulty reuniting the separations within in himself, for instance when dealing with ethical questions. He now has the opportunity to work on transcending the limitations when he leaves the Open University and moves to Schumacher College to work on reintegration. To a question from the chair as to whether it is by chance that this opportunity comes at the end of Goodwin's career, he answered that he doesn't regret the way he worked before and wants to recover the wholeness of vision without losing valuable qualities that have come out of the cultural experience.

Malnoë is working with disease resistance in genetically modified potatoes and was responsible for the first field trials in Switzerland. She was asked to described her first enthusiasm for genetics. She experienced that moment with very strong feelings in her first year of studying genetics when looking through a microscope for the first time she saw a cell division and its associated chromosomes. For her it was a precious, privileged moment of feeling one with the object she was observing. This experience has been the source of her drive in science ever since. It involves continuously trying to reach that state of mind again. The other aspect is that one gets so deeply into the research that one is no longer separate from it. Object and subject are one. She even described herself as addicted to science. Later on in her career she became involved with the release of transgenic plants into the environment and had to liaise with many interested parties from the public. This opened up her own views. It also made her realise that when doing that kind of research we are at the edge of knowledge, and that we don't know exactly what is going to happen in future. That is her dilemma. Plants involve a very complex ecosystem and we have to advance, but advance carefully, trusting that in doing so we shall identify the risks. She can cope with these dilemmas by having confidence in what earlier in this conference was called by Van der Wal: the primary world. Malnoë advised that we should proceed slowly, step by step so that we are able to oversee what is really happening.

For Hohn research is full of dilemmas. One is that science is not only research but also working with people. Science is a social process and it is a challenge to keep a group together. Another dilemma is that in research we specialise strongly and we reach a point where we know more and more about less and less until we know all about almost nothing. It is a problem always to maintain interest in other areas of research or life. Another of her dilemmas is being both a scientist and a mother, but she preferred this not to be discussed-

De Boer spoke of large and small dilemmas in science. One of them arose in his postdoc situation in a commercial company in the USA. After working enthusiastically on the expression of human growth hormone which is effective against dwarfism, the company asked him to work on bovine growth hormone (BST) which is known to stimulate milk production. Because of his farming background he didn't like this at all, but did it to earn money. This experience alerted him to the dilemmas which can face a scientist. His main drive is developing medicines based on proteins for the benefit of patients. Later on in his career he founded his own successful company (Gene Pharming) which works with transgenic dairy cows. His main concern was and is to set clear limits, at least for himself, by making sure that the health of animals is not affected by transgenesis. After he had left this company, it began to work on the expression of a very potent protein hormone involved in the development of red blood cells and to breed transgenic dairy cows to produce this hormone in the milk. But mice experiments had shown that they became terribly ill from the presence of this hormone. As founder and shareholder of this company he faced a very big dilemma. financially rewarding commercial interests were in conflict with his basic principles on animal health. In the end he publicly distanced himself from the company. The reason for this was that he wants to do work he can be proud of and feel able to tell his sons about. He would not be able to stand behind any research he felt ashamed of. When asked by the chair how important money is to him, he answered that money is very important along with a scientific interest in new research. For him it is the combination which is attractive. Although he considers himself first and foremost a scientist, he also finds it very interesting to start a business and attract money by making investors enthusiastic about gene technology. If he had the money he would leave the university research he is in now and start a commercial company tomorrow. His drive would be the commercialising of biotechnology by setting up production systems developing medicines through proteins.

Genetic engineering: a necessity?

Salgo explained that he came to the conference in order to reach a better judgement about whether he should take out patents on genetic engineering products or not. Out of personal conviction he is against patenting an onco mouse. To De Boer's statement, "No patents, no drugs – it is as simple as that", Salgo agreed that at the molecular biological level it probably is as simple as that, but he doubts if we really need genetic engineering. Malnoë doesn't think genetic engineering is an absolute necessity for the survival of humanity, but sees it as part of evolution and progress. She herself is working on two projects, namely viral and phytophora resistance in potato. Working on resistance in plants has an ecological aim in that it is an important way to reduce the amount of pesticide used. Genetic engineering is probably not the only way to make plants resistant to such diseases but it is much easier than traditional breeding. Through genetic engineering, plants can be obtained that are more efficient because there is less blending of different genomes. The traditional way has to go back to ancient varieties in order to find natural resistance genes which have been eliminated in breeding. By mixing these back into the modern varieties a lot of unknown phenomena are introduced, far more than when perhaps two gene sequences are introduced into a genome that is already established.

Hohn's work is much less applied than Malnoë's. It involves a combination of being driven by a kind of religious feeling for the work and facing many failures and disappointments. Doing basic research which involves wanting to know how something works, followed by the feelings of success at finding it out is what keeps her going. It is not important for her whether the research can be used, misused or not used at all. That is what the company she works for has to decide. For De Boer this kind of research is problematic because it is the company which gets the research reports first and can decide to use the results or not. At least if the work is patented others can read the details. Hohn admitted that she found the patent jargon difficult to follow. Goodwin asked if the sponsor had the right to decide whether or not to publish the results and whether one would want to work for a company which suppressed publication of results. Hohn asserted that her company has no right to prevent publication and that it can only slow it down. She would not work for a company that prevents publication if she had an alternative, but if there were no alternative she would rather do research than not do it! Publications are necessary for the young scientist's career. Hohn answered Goodwin's question with a specific real example showing that under certain circumstances suppression of information may be necessary because of risk to subsequent patentability.

Responsibility

To a question from the chair about who is to decide on the desirability of releasing genetically manipulated organisms such as potatoes into the environment, Malnoë stated that the scientist merely discovers what is possible, and that it is up to society then to decide whether or not to realise the possible. In Malnoë's opinion it is enough that a governmental committee of experts controls the research and its use, because she as the laboratory scientist does not have enough experience of the agricultural uses and consequences of her research. She added that of course she has her own opinion on her part of the work at laboratory stage. One scientist in the audience pointed out that if scientists insist on academic freedom for their work they also have to carry some of the responsibility for it. Malnoë agreed, but for her a decision based on responsibility must be achieved in a dialogue between different interested parties. However, she concurred with another questioner that if she had good reason she could of course withdraw what she is developing at any point in the process. She as the scientist cannot decide alone. Hohn, also a specialised laboratory scientist, supported this. De Boer added that the opinions of different specialists and interest groups have to be brought together by a committee and it is ultimately up to the Minister of Agriculture to decide. But of course the scientist should give all the available information. De Boer felt there is no reason why the products of classical breeding should not go through the same regulatory channel as genetic engineering products, because to him the risks are equivalent. Goodwin too said that the regulations should be as stringent for classically bred varieties as they are for transgenics. De Boer put this in a different perspective with the statement that the benefits of classical breeding over the last hundred years outweighed the risks. There were no accidents.

Future expectations

Pedroli's last question to the panel concerned what they wanted to have achieved ten years hence. De Boer wants to be at the birth of development of a medicinal product that could be used in a clinic to benefit patients. For Hohn it is a question of finding little pearls on a chain and wanting to come upon the biggest! For her, assuming that a virus can be regarded as living, a great achievement was the production of something that could live which was made in a test tube and could propagate in a bacteria. This was a great thing at the time and she hopes for more like it in the future. Salgo pointed out that he is sixty-five and thus does not have much longer to work, but he just wants to understand what a gene or a genome really is. Is it a blueprint, a building instruction, building information, or merely a carrier of information? Malnoë is just happy being part of a process of bringing a new technology into society. It is not always the easiest position, but for her it is very enriching to come into contact with people who have other opinions, as for instance at this open-minded conference.

Goodwin, also at sixty-five, likes to look forward too. What would give him much pleasure would be to look back in ten years time and recognize that all those present at the conference participated in a process of fundamental transformation of our view of nature and that these were the seeds that actually helped to bring about the necessary transformation.

27 Biotechnology as a socio-technical ensemble – closing remarks and reflection

GUIDO RUIVENKAMP
Sociologist
Working Group on Technology and Agrarian Development, Wageningen Agricultural University
Nieuwe Kanaal 11, NL-6708 PA Wageningen

Introduction

The organisation of the conference has been successful in bringing persons together from different disciplinarian backgrounds, willing to listen to each other and to learn from each other. The presentations were of a high quality, introduced by Edith Lammerts van Bueren, as chairman, in such way that it became easier for the audience to follow the general line of argumentation.

Indeed, an intensive discussion took place about the presuppositions in science and expectations in society. Evaluating the conference three important contributions and two shortcomings are mentioned below.

Criticising the reductionistic scientific approach

Different speakers emphasized the need to develop another scientific approach with more respect for the intrinsic value and potential of nature (Meyer-Abich, K.). Besides, there was a general plea for another style of scientist. The scientist as a cool observer was criticised and one emphasized the need for a scientist as a participatory perceiver.

In the specific field of biotechnology, the conference stimulated the reflections on the DNA-concept. It was wondered if the DNA-concept was not primarily a cultural concept (Lindee, S.). It was said that it was a fuzzy concept, even an unscientific concept (Fischer, E.P.). Instead of speaking about genes you had to speak about nuon, protonuon, maptonuon and xaptonuon (Brosius, J., Gould J, quoted by Fischer E.P). Nobody really knows what a gene is, and because it is such a fuzzy concept everybody talks about it (Fischer E.P.). Finally, ideas were presented to differentiate between old and new genetics; between a very instrumental, reductionistic versus a more holistic approach (Ho M.W.). Also, in relation to the DNA-concept, it was emphasized that more attention should be paid to the contextuality; that genes function in a complex non-functional network, that genes and genomes are not stable but dynamic (Ho M.W.).

The general plea of the conference was to stimulate scientists to adapt a more holistic as well as a participatory approach.

Criticising the separation of science and society

In different sessions the organisers have also tried to go beyond the image of the pure scientist. The conference stimulated the participants to pass by their disciplinarian frontiers and to explain in which way personal considerations were linked to scientific activities. Indeed, the participants increasingly explained in which way they weighted ethical and scientific elements in their work. This discussion about the relation of these two worlds, the world of science and the world of life, was the central topic of the very stimulating presentation of Jaap van der Wal and has had a big impact on the conference. It became clear that the image of a separation between the pure (rational) scientist and person was no longer realistic. Personal affairs, ethical considerations influenced by debates in society, played a role in the individual choices of the scientist.

Criticising the separation of technology and society

During the conference it was often asked why the technologies and especially biotechnologies are developed as they are. Some negative examples of the impacts of new biotechnologies were mentioned, such as substitution of agricultural products. But no answer could be given to the question if for example participatory biotechnology is possible. The presentations criticize the gap between the demands of society and the actual biotechnological developments but there was no clear comprehension of the interwovenness of biotechnological and societal developments. This interwovenness was neither a topic of discussion. This lack of attention for the relationship between technology and society is, in my opinion, linked to the fact that – despite all efforts – a reductionistic approach was still dominant at the conference.

This reductionistic approach became manifest in the following two characteristics of the *If*gene conference.

Scientific research a topic in itself

Although the presentations about the DNA-concept, new genetics, etc. were very interesting, the impression was also raised that scientists could choose freely in following one or another scientific approach. The discussion focused on scientific dilemmas: old or new genetics (Ho, M.W.); to appropriate nature or to increase the potential of nature (Meyer-Abich). These alternative scientific approaches were presented as being equally accessible for the individual scientist. The illusion was made that choosing between these approaches was a pure scientific issue, Indeed, no analysis was presented about the socio-economic pressures to work with specific scientific notions. The impression was raised that scientist could easily change from an instrumentalistic to a more holistic approach; as if specific scientific developments were not embedded in broader socio-economic networks with large interests. Therefore, the discussion remained "pure" scientific and often "irrealistic".

Scientist as a free individual
Another characteristic of the conference was the intention to show that the participants were people, with scientific as well as with ethical and personal considerations (see point b). Although this approach was interesting, it also reinforced the image that scientists can choose freely; that there is no social control; no market mechanism.

By paying so much attention to personal ethical weighting the conference gradually ignored that these individual choices take place within society, characterised by unequal power relations. Indeed, no discussion took place why – despite the personal preferences of an ecological agriculture – most scientists are still (obliged) working at the development of e.g. herbicide-resistant crops. It was neither discussed why the main biotechnology research programmes are focused on the major food crops of the industrialized world; why primarily those products, come on the market (long-tenured tomatoes, BST) which suit the interest of multinational corporations and which are adapted to an industrialized agriculture. Because of this emphasis on ethical, individual choices of the scientist, society and laboratory were still considered as being separate units. That's why in the following paragraph I want to indicate shortly in which way biotechnological developments are strongly related to the social organisation of the agro-industrial chain of production. This interwovenness leads to a *specific* development of biotechnology. This increases the responsibility of scientist to choose, but socio-economic pressures will make it very complicated to take that responsibility.

Agrobiotechnology and its contextuality
The development and application of biotechnology does not take place in a historical vacuum, but reflects existing economic and political relations of power. Indeed, some groups of actors (for example transnational corporations) have more influence on the directions of the scientific research than other groups (for example farmers in third world countries). Therefore, the discussion on the developments of biotechnologies should not be characterised by a philosophical discussion on possible applications of a "socially neutral" technological development.
Instead, it should focus on the relation between existing developments in the agro-industrial chain of production and the resulting specific development of the biotechnological potential.
Therefore, it is necessary to take first into consideration that agriculture can no longer be considered as an independent sector. Agriculture should be viewed as one phase in the agro-industrial chain of production. This implies that the situation of agriculture is determined by its interactions with companies that supply agricultural inputs as seeds

and agrochemicals and with companies which transform the agricultural output into food products. The development of biotechnology takes place within this context of the food chain in which specific patterns of relations exist between the different actors.

Developments in the first phase of the agro-industrial chain of production

The first phase of the agro-industrial chain of production is characterised by the following three developments:

a) *Concentration* of transnationals (TNC's) in the production of agrochemicals and seeds: A limited number of TNC's control largely the market of pesticides. A major part of the international seeds sector is now being controlled by some 20 TNC's, most of which are also major pesticide producers.

b) Concentration of agricultural production of a few crops and *a few varieties of each crop*. It is important to keep in mind that despite the enormous diversity of agricultural products, only thirty crops account for 95% of the world food consumption. Moreover, only three crops (wheat, corn and rice) provide more than half of all foodstuff.

c) Integration of the seeds sector and that of pesticides. Therefore, we can speak of the birth of a new industrial sector: the *genetics supply industry*.

The limited base of the food-package, taken together with the narrowing genetic basis for the most important crops (uniform varieties), determines the political significance of the introduction of biotechnology into the genetics supply sector.

This industry is composed of companies which have one foot in pesticides, and another in seeds and genetic engineering. Biotechnological research by the genetics supply industry is not so much done to replace seeds and/or agrochemicals, but first of all to modify these seeds and agrochemicals, to change the highly specialized packet of genetic information, which controls how the plant will grow and respond to its environment. The possibility of intervening in this program implicates that one can determine where, when and how the growers will sow, harvest and care for their crops. Companies which control these inputs, are well on their way to control food production at a distance. Scientist working at these new products are contributing to a social reorganization of the food production.

Developments in the third phase of the agro-industrial chain of production

Agricultural products are transformed by industry into food products and agriculture becomes increasingly a subsector of that industry. This development to integrate agriculture with food processing implies that technological modifications in food processing will have a big impact on socio-economic developments in agriculture. Therefore, it should be asked how biotechnology will change the interaction between agriculture and the food processing industry.

Four forms can be differentiated in the processing of agricultural products:

a) To conserve agricultural products,
b) To transform agricultural products,
c) To combine agricultural products into new food products,
d) To split up agricultural products into components which may be used in a broad spectrum of final food products (and/or non-food products).

Although all those forms of processing techniques continue to exist, biotechnology seems to speed up the processing technique of splitting up and reaggregating food components from different agricultural crops.

The social dimension of biotechnology in the third phase of food production goes through the development of better/new catalysts, such as enzymes, which extract nutrients from a broad range of agricultural crops. This implies that the linear organization of food production, from a specific agricultural product to a specific food product, collapses.
Basic nutrients as carbohydrates, proteins, fats and elements as amino acids or vitamins are extracted from a broad spectrum of agricultural crops and/or manufactured in industrial fermentation processes.

This means that different crops like sugar and corn (high fructose corn syrup) become parts of a new "sweeteners sector" and that crops such as wheat, soya and (again) corn can also become part of a new "protein sector".

The interchangeability of agricultural products (for example corn and sugar) as well as the fermentative production of food components (for example aspartame) will free the food processing industry from necessary supplies of specific agricultural products (for example cane-sugar).

The separation of food- and agricultural production implies a reorganization of the food chain in which the interchangeability of products becomes a central issue. Companies which develop the scientific and industrial capacity to reprogram microorganism and their enzymatic activities to extract food components from different commodities, have in their hands the mechanisms to divide and control the producers – farmers and workers – of these basic nutrients. Scientist working at the development of enzymes and microbiological products are involved in realizing an increased interchangeability of agricultural products.

Biotechnology for an industrializing agriculture

The actual global production system of food is tied up between inputs like seeds, fertilizers, pesticides machines and satellite information on the one hand and post-harvesting technologies like processing, storing, exports and marketing on the other. The development of biotechnology is politically important, because research and applications concentrate on the strategic links in food production, such as seeds, agrochemicals, enzymes and amino-acids. Through the deliverance of new products (herbicide-resistant crops, enzymes) the relations between the actors within the food chain are influenced. Biotechnology contributes to reorganize the relations between farmers and the companies which deliver the inputs as well as the relation between farmers and food processing industries.

The analysis of the development of biotechnology in the different agricultural branches and phases of the agro-industrial chain of production have made clear that current biotechnological developments are especially related to the interests of industry. Because of this specific social dimension one can speak of "industrial biotechnology". This "industrial biotechnology" promotes a specific form of integration of the agricultural sector within the agro-industrial production chain. Thereby a new controlsystem of agricultural production and farm labour is created. This reorganisation process of the agro-industrial chain of production, which is characterised by:

a. A growing control, at a distance, over the agricultural sector by industrial companies through the introduction of new inputs (for example, seed, pest-control methods and breeding material);

b. A growing interchangeability of agricultural products, production areas and methods;

c. Increased privatisation of the crucial links of the food chain through patents which reinforce the political position of the private companies and weaken the political position of public and social organisations; like unions, farmers and non-governmental organisations.

This specific use of biotechnology clarifies the existence of "the acceptability problem" for different social groups concerning "industrial biotechnology". It also illustrates the need to change the social character of biotechnology and to relate the social-economic content of biotechnology also to the interests of other stakeholders. The condition for this is, that the development of biotechnology is related to another social context than that of industrialized agriculture within global food chains. To link biotechnologies to other contexts, such as the contexts of sustainable and diversified farming systems, imply another kind of professionalism for the plant breeders, enzyme technologists and microbiologists. They have to learn to integrate social and technical dimensions in their work; to find out possibilities to tailor biotechnologies to wider social benefits; to take their social responsibilities in their laboratories.

Agrobiotechnologies for wider social benefits
Recombinant bovine growth hormone, herbicide- and pest-resistant crop varieties, long-tenured tomato, etc. illustrate that biotechnology has been used in this context of industrial agriculture of global food chains. All these innovations refer to the demand of an increased yield and conservation of agricultural bulk products. However, it is not inevitable that biotechnology refers to this context it might also be related to sustainable farming practices.

Currently, still relatively little amount of work has been done to explore the possibilities of tailoring biotechnology to diverse and sustainable farming systems. Therefore, it can be said that *modern biotechnology presents a paradox.* In principle it offers wide technical possibilities for increasing food security and agro-biodiversity. The reality is, however, that primarily those applications are pursued which have only sense in the narrow social frame of an industrializing and uniforming agriculture. This narrow social context, in which biotechnology is evolving, needs to be questioned. Educational and research programmes are often part of that narrow social context; "are part of the problem". Therefore, new initiatives in collaborative research and education programmes are needed in the design of agrobiotechnologies for wider social benefits (note 2).

New professionalism
Biotechnology is not merely a group of techniques, it also contains and represents social dimensions. Considering biotechnology as a *hybrid socio/technical science,* new challenges emerge for natural scientists, social scientists and for members of non-governmental organisations (NGO's). For all these practitioners it becomes imperative to explicitly reflect on the social embodiment of biotechnology and to link concretely its social and technical dimensions. The spread of new agricultural biotechnologies and their impacts on the social relations of production in rural economies have already raised a growing demand of different social groups (consumers, farmers, environmentalists) to co-steer biotechnological developments. No longer can new biotechnologies be generated and implementated from a top-down approach. Neither can they be developed from a strictly disciplinarian approach. Biotechnological developments will increasingly be *negotiated.*

This requires a shift in educational and research practices. Educational and research programmes can no longer be exclusively focused on either developing techniques or assessing the social impacts of these techniques. Ultimately, these programmes should aim to overcome the separation between natural/technical and social/economic studies. Besides, social and technical scientists should learn to work together with other stakeholders (farmers, consumers, policymakers) in the development of biotechnologies. Finally, a new social debate should be organized how to develop and apply bio-technologies in the food-chain.

Limitations in the actual biotechnology-debate

In the actual debate we can differentiate three ways of discussing biotechnological developments. Broadly speaking there are three groups of participants.

Splitters

These persons separate biotechnology from its social context. The splitters consider biotechnology as just a group of techniques. Whenever, they discuss the social dimensions of biotechnologies they refer to the consequences of the technologies for society. The splitters are divided in two subgroups, the optimist and the pessimist ones; those who think that biotechnology might resolve all problems (the supporters of biotech) and those who emphasize the negative impacts of biotechnology. However, supporters and opponents of biotechnology start from the same basic idea that it is possible to discuss the impacts of factor A (biotechnology) on factor B (society).

By discussing the positive or negative impacts of biotechnologies the splitters are not questioning why the products are developed as they are. And if they do ask it, than they emphasize that biotechnological developments are the results of free choices of the individual laboratory researchers.

The splitters refer to the linear model of scientific development. Science is developed in the laboratory and provides rules that are used in technology and the technology is applied in society and creates changes in society.

Because of their thinking in cause and effect, the splitters are very active in the social debate and ask for a social evaluation of biotechnology as a group of techniques, which influences society. By separating biotechnology from its context, the opponents and supporters of biotechnological developments can agree upon a case-by case evaluation. They are also inclined to differentiate between good and bad biotechnologies. Besides, the splitters are especially interested in the most advanced examples of the new techniques. Indeed, the spitters incline to focus on the most spectacular examples of biotechnologies on which they base their judgements.

Weavers

The weavers emphasize that biotechnology does not cause social change but that it is an integral element of that process. They emphasize that there is no hammer without a hand, and that the hand is under social control. They refer to the social organization of labour. Starting point is the close interrelationship between biotechnological and societal developments. The weavers stress that biotechnological developments are not accidental but that they reflect social developments. Weavers are interested in which way characteristics of the actual social organization of the food chain are incorporated in the biotechnological developments. Indeed, for analysing biotechnology the weavers start with analysing society. Biotechnologies are no longer considered as a group of techniques but as an ensemble of social and technical dimensions.

The splitters have often problems to understand the weavers. The splitters want to focus on the concrete impacts of factor A (biotechnology) on factor B (society) and therefore often consider the weavers as being abstract or vague. They also argue that the weavers do not differentiate; that their accusations on biotechnology are not correct, mainly because weavers refer to general (societal) developments. The weavers understand that criticism of the splitters, sometimes even agree with it, but emphasize that they are exactly interested in the interwovenness of biotechnological and societal developments.

Redesigners

The redesigners agree with the weavers about the interwovenness of biotechnological and societal developments, but they aim to change that relationship. The redesigners are the optimists searching for room of manoeuvre in the actual soc/technical webs.

The redesigners criticize the weavers that they only describe the interwovenness and they do not accept a social deterministic point of view, often present by the weavers. The redesigners argue that it is not inevitable that actual biotechnological developments are in the interests of TNC's. Biotechnologies can also be placed in other networks, related to the interests of other groups. The redesigners are looking for possibilities to relate biotechnological developments explicitly to broader social interests. The redesigners are networkers trying to find colleagues to work at changes within different institutes.

Conclusions

The conference emphasized the need for a more holistic and participatory approach. This is easier said than done. The first step is to broaden the scientific way of thinking so that it is more in accordance with contextual biology. But a reductionistic approach still remains in the sense that there is a biological context (science) which continues to be separarted from the socio-economic context; thinking in terms of cause and effect as well as believing in the existence of free intellectuals is still widespread. Coming to another approach and dealing with biotechnologies from a redesign perspective implies many changes which will often be difficult to realize. First, the criticism on the instrumental reductionistic approach in all related fields should be more developed. New educational and research programmes will be needed. Funds for transdisciplinarian and participatorian might be difficult to find in a society built up along disciplinarian frontiers. Secondly, scientists should become aware that – through their laboratory activities – they contribute to changes in the social organization of food production. If scientists are prepared to consider themselves as social-engineers, embedded in a specific socio-economic environment and working within processes of social changes, then it can be asked if scientists are willing to reflect on their socio-technical choices.

It is often "forgottèn" that every technology – and therefore also biotechnology – can be developed in different ways. Breeding techniques can be applied to make new varieties pesticide or pest resistant. New products can be developed (such as productive varieties and animals) which prescribe one specific way of farming; products can also be developed to meet the differing styles of farming and thereby strengthen the heterogeneity of the agricultural sector. Scientists can work for biotechnologies, related to an uniform and industrialized agriculture or they can work for biotechnologies tailored to sustainable and diversified farming practices. In other words, the development and application of biotechnology into agricultural production should be negotiated. It might be a good follow-up for this conference to exchange information about the possibilities of relating biotechnologies to sustainable and diversified developments.

The efforts during this conference to broaden views relating to the biological and human contexts were experienced as a new and beneficial initial step, but the next step should extend to the socio-economic context of biotechnology!

Notes

Although I present here only my critical comments, I want to stress that in general I appreciate very much the discussions on the conference.

The working group on technology and agrarian development (TAD) of the Wageningen agricultural university is trying to form networks in these fields. The objective is to develop collaborative research and new educational programmes (including new Msc. courses). Persons, interested in participating in these topics are asked to write to TAD.

28 *If*gene network

*If*gene is an international network of people from various disciplines who have questions relating to gene technology and are engaged in the process of forming judgements about it from as broad a perspective as possible. The *If*gene conference The future of DNA was a crystallisation point in this process.

*If*gene is an initiative of the Naturwissenschaftlichen Sektion der freien Hochschule für Geisteswissenschaft at the Goetheanum in Dornach (Switzerland).
Within the network of *If*gene working groups have already been established in several countries. For further information about meetings of these groups in your region please contact your nearest coordinator:

France
Christine Ballivet
Institut Kepler
6, av. Georges Clémenceau
F-69230 St. Genis Laval
Tel: ** 33 7856 1941
Fax: ** 33 7856 8457

Germany
Meinhard Simon
Limnologisches Institut
UNI Konstanz
Postfach 5560, D-78434 Konstanz
Tel: ** 49 7531 8831 12
Fax: ** 49 7531 8835 33
Email:meinhard.simon@uni-konstanz.de

Manfred Schleyer
Karl-Ernst von Baer Institut
Im Kreuz, Am Marktplatz
D-87764 Legau
Tel./Fax: ** 49 8330 1544
Email: MSchleyer@aol.com

Great Britain
David Heaf
Hafan, Cae Llwyd,
Llanystumdwy, Cricieth
Gwynedd, LL52 OSG Wales
Tel: ** 44 1766 523181
Fax: ** 44 1766 523181
Email: 101622.2773@CompuServe.com

_If_ gene is also on Internet.
Home page at: http://www.peak.org/~armstroj

Netherlands
Edith Lammerts van Bueren,
Louis Bolk Instituut/ Werkgroep Genenmanipulatie en Oordeelsvorming
Hoofdstraat 24
NL-3972 LA Driebergen,
Tel: ** 31 343 517814
Fax: ** 31 343 515611
Email: louis.bolk@pobox.ruu.nl

Switzerland
Johannes Wirz
Naturwissenschaftliche Sektion am Goetheanum
CH-4143 Dornach, Switzerland
Tel: ** 41 61 706 4210
Fax: ** 41 61 706 4215
Email: 100716.1756@CompuServe.Com

USA
John Armstrong
25126 Pleasant Hill
USA-Corvallis, OR 97333
Tel: ** 1 541 929 3045
Fax: ** 1 541 754 6641
Email: armstroj@peak.org

Subject index

MIX
Papier aus verantwortungsvollen Quellen
Paper from responsible sources
FSC® C105338

If you have any concerns about our products,
you can contact us on
ProductSafety@springernature.com

In case Publisher is established outside the EU,
the EU authorized representative is:
Springer Nature Customer Service Center GmbH
Europaplatz 3, 69115 Heidelberg, Germany

Printed by Libri Plureos GmbH
in Hamburg, Germany